SKYWATCH
THE WESTERN WEATHER GUIDE

SKYWATCH
THE WESTERN WEATHER GUIDE

RICHARD A. KEEN

FULCRUM INC
GOLDEN COLORADO
1987

Cover Photograph by
Tom Ives
©1987

Cover and Book Design by
Chris Bierwirth

Library of Congress Catalog-in-Publication Data

Keen, Richard A.
Skywatch: the western weather guide.
Bibliography: p.
Includes index.
1. West (U.S.) — Climate. I. Title
QC984.W38K44 1987. 551.6978 87-11871

ISBN 1-55591-019-X

printed in Hong Kong
through Interprint, San Francisco

cover photo: lightning storm over Tucson, Arizona

To Mom and Dad
Thanks for reading the rain gauge all these years.

CONTENTS

ACKNOWLEDGMENTS

I must confess from the start that writing this book wasn't really my idea! For years I have been giving lectures to the Colorado Mountain Club about the strange and spectacular weather that hikers and mountaineers might encounter, and the idea for this book actually came from the audience one evening—in particular, from Fulcrum's Betsy Armstrong (co-author of *The Avalanche Book*). Betsy thought that the weather of the West would make a lively subject for a book, and her contagious enthusiasm got the project rolling.

Along the way, the folks at Fulcrum, especially Hunter Holloway and Dr. Archie Kahan, gave many useful comments and suggestions that greatly improved the final text. Thanks are also due to dozens of fellow weather buffs—old friends and new acquaintances alike—who provided all sorts of assistance. Among these, Dave Blanchard, Ron Holle, Dennis Rodgers and others of the National Oceanic and Atmospheric Administration (NOAA) were especially helpful. Other organizations whose people contributed in one way or another are the National Center for Atmospheric Research (NCAR), National Weather Service, U.S. Forest Service, Federal Emergency Management Agency, the Universities of Chicago, Colorado and Hawaii, Colorado State University, Western Oregon State College and the Boeing Company.

A very special appreciation goes to Helen Duran for her encouragement and good humor during the long hours of writing.

A wild lady named Hazel also deserves mention. When this great storm visited my hometown in Pennsylvania one October evening in 1954, she brought misery to millions. To at least one wide-eyed seven-year-old, however, she brought a sense of wonder and lasting fascination. In a way, that stormy night thirty-three years ago is where this book really began.

Richard A. Keen
May 1987

"Frankly, I don't like the look of the weather . . ."

INTRODUCTION

To many, the West conjures up images of rugged scenery and even more rugged souls who spent their lives trying to tame it. Names like John Wesley Powell, Geronimo, Wyatt Earp and John Wayne have become legends around the world, and millions cross oceans and continents to visit places like Pikes Peak, Death Valley and the Grand Canyon. With the eruption of Mount Saint Helens, we have even seen the capricious landscape explode in our faces! So it is fitting that the weather of the West is every bit as awesome and fickle as the land it sweeps.

Those who live in the eleven western states are already familiar with the power of their weather. Westerners have seen winds strong enough to send parked airplanes into unwanted flight and airborne planes crashing to the ground, and tempests that have destroyed bridges and flattened trees by the millions. Coastal storms have washed neighborhoods into the sea, while trains have been stranded and houses buried in mountain snows. And desert whirlwinds have sent terrified jackrabbits sailing skyward!

The West can get hot—only the Sahara has recorded higher temperatures than Death Valley. Montana and Utah have experienced some of the coldest weather ever recorded outside the Arctic and Antarctic. Between the extremes, the temperature has been known to fall 100 degrees overnight. Yet, for the most part, the climate of the West is pleasant. Sunshine is abundant over most of the region, and those few cloudy areas are compensated by moderate temperatures.

Remarkable changes occur in western weather from one day to the next, from summer to winter and from year to year. Changeable weather is common in the middle latitudes, and the old quip, "If you don't like the weather, wait a minute," is invoked in many languages in many places. In the West, though, one can also add, "If you still don't like the weather, go a few miles." An hour's drive will take many westerners from forest to desert, or from summer heat to a snowfield.

This variety hasn't gone unnoticed. Many segments of the economy—from agriculture to recreation—thrive on it. Tropical fruits are grown with water drawn from nearby snow-clad mountains and, in return, the mountains receive hordes of skiers from warm cities in the lowlands. Yet, these same activities can also suffer from the varied weather—frostbitten oranges and snowless ski runs are among the many reminders of the whims of western weather.

What makes the West's weather the benevolent prankster that it is? It's a simple question with many answers, and those answers are what this book is all about. Descriptions of the fascinating variety of western weather—through words, pictures and numbers—make great reading. But this book intends to go beyond that, I hope, and leave the reader with an understanding or, better yet, a *feeling* for how the West's peculiar weather works. This is where the real satisfaction lies. It's like looking under the hood of a car and knowing why the distributor is there, or realizing how a human heart does its job.

Some reasons for the West's unique climate are obvious. It's a region of mountains and valleys with the world's largest ocean next door. But for a real explanation, we must go far beyond local topography. Shifting currents in the equatorial oceans and the frozen wastes of the North and South poles are every bit as important to western weather as are the Rockies and the Sierras. The importance of all these factors, however, is dwarfed by the role played by ancient volcanoes that have since eroded into sand. They gave us air, and without air there would be nothing to write this book about. The story of these 4-billion-year-old eruptions unfolds in the next chapter.

It may seem strange that the second photograph in this book is a landscape from the planet Venus, but there's a real lesson to be learned from our sister planet. Over the eons since those primordial volcanic eruptions, earth slowly became the garden it is today—while Venus became a searing visage of hell. If nothing else, a look at the different fates of the two planets should help us appreciate how truly fortunate the earth and its inhabitants are. Think about it the next time you see lightning cracking the skies over the Grand Canyon. It's a sight you'll see nowhere else on earth or—barring some parallel solar system out of a "Star Trek" episode—in the entire universe!

Admittedly, western weather is far too complex to be stuffed into a book. Like many things in life, the best way to learn about the weather is to observe it in detail. Tips on tracking the daily passage of foul and fair weather thus conclude this book with a chapter on "Watching the Weather." I hope that your learning experience has just begun when you put this book down.

Between the creation of the atmosphere and the construction of your personal weather station, the book progresses in what seems to me an orderly and logical manner. The chapters are meant to be read as independently of each other as possible. So if you have a hankering to read about hurricanes, you can do so without having to flip through the entire book.

Finally, a word about metric versus English units of measurement. I've always felt that the main purpose of a measurement system is to communicate information. Since the odds are that most readers of this book are more familiar with the English system (or American system, as it's known in some remote valleys), English units it is! Snow and rain fall in inches, hailstones are weighed in pounds, wind blows in miles per hour and all temperatures are in degrees Fahrenheit. True, it's an old system of measurement and its arithmetic gets a bit tricky (quick: how many inches in a mile?), but as long as we both know how deep a foot of snow is, it works quite well.

WHY IS THERE WEATHER?

Why is there weather? The weather may seem such an ordinary ingredient of our daily lives that we easily forget that it, too, has a purpose. Of course, on a snowy morning it may seem obvious that this is simply to make life a little more miserable. But not only can these nasty storms make life miserable, they also—in a very real sense—make life as we know it possible. Weather, that alternation of sun and storm and of warm and cold, is inextricably intertwined with Earth's global ecosystem, and without it the world would be a vastly different place.

Five basic factors combine to make it inevitable that the Earth has the kind of weather that we have grown accustomed to. The first, and most obvious factor, is that Earth has an atmosphere. Second, Earth is sunlit. The third factor is Earth's rotation. The next factor—and one unique to Earth—is our planet's vast supply of liquid water. And finally, there is geography—the variety of surfaces, from oceans to continents to ice sheets, that cover Earth. In recent years we have been treated to close looks at other planets in our solar system, where the factors that control what might be called "weather" are incredibly different from ours. A survey of the conditions leading to Earth's weather will help us understand why it is the way it is, and comparisons with other planets will let us appreciate its uniqueness.

THE ATMOSPHERE

Earth has about five million billion tons of air. That's a lot of air; numbers like this make even the national debt seem small. But there are always ways to make small numbers look big and big numbers look small. To make the bulk of the atmosphere seem small, consider this: all those millions of billions of tons of air amount to a mere millionth of the weight of the entire Earth. Another way to look at it is to imagine a winter cold enough that the air itself freezes and falls to the ground as a layer of solid nitrogen and oxygen.

HOW TO MAKE WEATHER

Give your planet
some air

Heat it with sunlight

Rotate the planet to get
the air currents turning

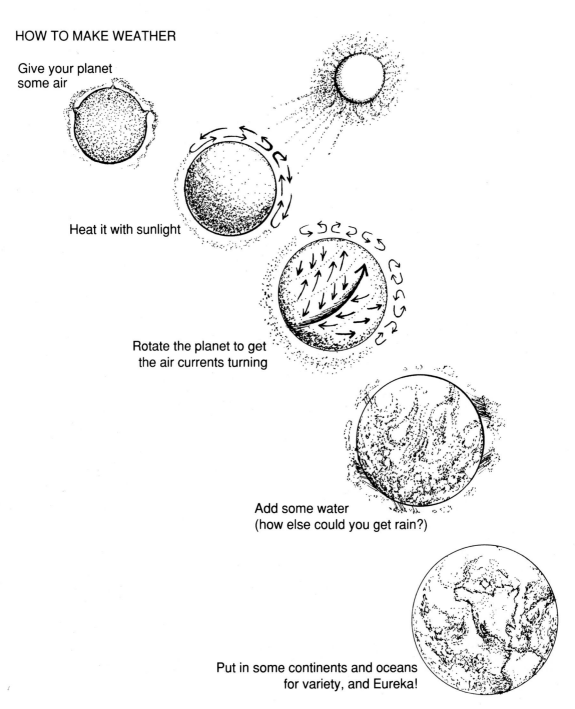

Add some water
(how else could you get rain?)

Put in some continents and oceans
for variety, and Eureka!

NASA

EARTH—Combine the five ingredients in the proper proportions and the result is the weather of Earth.

This layer would be about 20 feet thick. That's a lot of frozen air to shovel your car out of. But on a world 8,000 miles wide, 20 feet isn't really a whole lot, is it? Anyway, that's how much air there is; you decide whether it's a lot or a little.

Where did the atmosphere come from? Four and a half billion years ago, when the planets formed out of dust and gas surrounding the new-born sun, leftover gases collected around the in-

fant planets. These gases were mostly hydrogen and helium, which is not surprising since they are the most abundant elements in the universe. But hydrogen and helium are also the lightest of the elements, and Earth's gravity—strong as it may seem to us—was too feeble to keep them around for long. Farther from the sun, the gases were cooler and therefore more able to stick around the young planets. To this day, the distant planets

5

NASA

VENUS—This view of the surface of Venus, drawn by NASA artists, is a scene straight out of Dante's Inferno. *In size, Venus is Earth's twin sister, and its early days were probably much like Earth's. Along the way, however, Venus evolved into a parched hell, with a thick 850° atmosphere. Life as we know it is completely impossible on Venus.*

Jupiter, Saturn, Uranus and Neptune still have massive atmospheres of hydrogen and helium. Earth, however, like Mars, Venus and Mercury, soon found itself without an atmosphere.

What Earth lost to space, it soon (geologically speaking) created from within. Heat in Earth's core sent columns of molten rock to the surface, erupting as volcanoes the likes of which we don't see very often anymore. Substances trapped in the rocks, such as water,[1] carbon dioxide, methane, sulfur and nitrogen, spewed into the vacuum surrounding Earth. Unlike hydrogen and helium, these heavier gases stayed close to the Earth's surface. Thus, our atmosphere was born.

By today's standards, this early atmosphere was a vile affair, consisting largely of carbon dioxide, steam and clouds of sulfurous gases and with no breathable oxygen. But a little chemistry and, later, biology took care of the oxygen shortage. Most of the water fell as rain to form the oceans. For the first billion years or so, carbon dioxide mixed with water vapor to make rainfalls of carbonic acid. This acidic rain leached calcium from the surface rocks, and the ensuing chemical reactions formed calcium carbonate. Streams and rivers carried the calcium carbonate into oceans where it sank to the bottom, forming the thick limestone layers we see scattered around the globe.

It is interesting to compare Earth's history with that of Venus, 26 million miles closer to the sun. Although Venus is similar to Earth in size and (probably) composition of volcanic gases, its atmosphere is a hot and dense soup of mostly carbon dioxide. Apparently, intense sunlight on Venus prevented the planet from cooling quite enough for water vapor to condense into liquid. Without liquid water, no rains of carbonic acid could fall onto the rocks below. Venus' carbon dioxide remains in its atmosphere, while Earth's is locked in underground limestone, and all because of the relatively slight difference in distance from the sun. One may speculate on how narrowly Earth escaped Venus' hellish fate, or how narrowly Venus missed becoming the paradise that is Earth.

Half the diameter of Earth and one-tenth its bulk, the little planet Mars never had enough

MARS—More familiar perhaps, is this view of a Martian landscape, taken by NASA's Viking lander in 1976. The scene could be Death Valley, California, or Sunset Crater, Arizona, but it's not—it's Mars. The weather on Mars might also seem familiar, with clouds, frost, snow, fronts and dust devils. However, Mars' thin atmosphere is less than 1 percent as dense as Earth's, and even in the Martian tropics, nighttime temperatures regularly reach 125 degrees below zero. With no liquid water, Mars is an eternal desert that is, so far as we know, lifeless.

NASA

volcanoes to create much of an atmosphere. Like that of Venus, Mars' air is mostly carbon dioxide, but it is incredibly thin—only 1 percent as dense as the Earth's, and a mere ten-thousandth the density of Venus'. Nonetheless, thanks to trace amounts of water vapor and a rotation rate nearly identical to Earth's, Mars has more Earth-like weather than any other place we know of in the universe. Automated weather stations sent to the red planet have detected fronts, cyclones, clouds and even snow. But Mars is still a desert planet, and, so far as we know, lifeless.

Meanwhile, back on Earth, at the tender age of one billion years, plant life began releasing oxygen (through photosynthesis) into the atmosphere. Eventually, about half a billion years ago, there was enough oxygen to allow a layer of ozone (a molecule of three oxygen atoms) to form. Ozone constitutes only the tiniest fraction of the atmosphere—less than one molecule in a million. However, this minuscule amount is extremely important because it absorbs the sun's ultraviolet radiation, helping to protect us from sunburn and skin cancer. (This beneficial shield of ozone, 10 to 20 miles high, shouldn't be confused with local concentrations of ozone that form around today's cities.) Protected from ultraviolet rays, early plant life increased dramatically. The greening of Earth led to a rapid increase in the amount of atmospheric oxygen, leading in turn to oxygen-breathing animals like dinosaurs, who needed lots of oxygen and, at last, us. The modern-day status of the atmosphere is comprised of 76 percent nitrogen, 23 percent oxygen and 1 percent other gases.

SUNLIGHT

All this air would simply sit still without some source of energy to move it around. That source of energy is the sun. It may seem obvious to us that the sun is the main source of heat for the atmosphere, but there are planets where that isn't the case. On Jupiter, the largest of the planets and

five times Earth's distance from the sun, sunbeams are so feeble and the planet's center so hot that the atmosphere is actually heated more from below than above. On Earth, though, most of our heat comes from the sun.

Not all places on Earth get the same amount of heat. In the tropics, when the noontime sun passes directly overhead, the ground gets the full effect of solar heating: a one-foot-wide beam of sunlight heats up a square foot of ground. At higher latitudes outside the tropics, the noontime sun doesn't pass overhead, and its light strikes the ground at some oblique angle. Forty-five degrees from the equator, the foot-wide sunbeam has to heat one and a half square feet of ground, and at 60 degrees latitude the sunbeam is spread over two square feet. At the poles, 90 degrees from the equator, the sunbeam hits the ground at such a low angle that it is spread over many square yards. Accordingly, the equator is heated more by the sun than are the poles.

As the sun heats Earth in its non-uniform manner, the planet warms up. If this heating went on forever, the planet would eventually melt. Since Earth as a whole is *not* heating up, there must be some way for the planet to get rid of the solar heat as fast as it receives it. Earth does this by sending energy out to space in the form of infrared radiation. *Infra-red* is a mixture of Latin and English meaning 'below red,' and refers to light waves whose frequency is less than that of red light. Although we cannot see infrared radiation, there are times when we can feel it—like the warmth we feel coming from a hot stove or from a radiator (hence the name!).

On the average, Earth must lose as much heat through radiation as it receives from the sun. As with sunlight, this loss of heat is not uniform around the globe. A hot object radiates heat faster than does a cold one; at a temperature of 80 degrees, for instance, the tropics radiate heat twice as fast as does the zero-degree Fahrenheit Arctic.

So not only do the tropics receive more heat than do the poles, they also lose it faster. However (and this is very important), gains and losses do not balance out locally, even though they do balance for the globe as a whole. In the tropics, there is a net gain of heat, as the incoming sunlight is greater than the loss through radiation. Conversely, the arctic regions see a net loss of heat. It is this difference in the *net* heating between the equator and the poles that drives Earth's weather.

Were the net heating of the tropics and the net cooling of the arctic regions to continue unhindered, Earth would end up with three equally obnoxious climate regions—an unbearably hot equatorial zone bounded on either side by two frightfully cold polar areas. The atmosphere, however, keeps this from happening. Remember that hot air is lighter than cold air, so as the air over the equator heats up, it rises. Conversely, the cooling of air over the Arctic causes it to sink. The rising equatorial air moves toward the poles, replacing the polar air that has sunk to the ground and is now crawling towards the equator. In this manner, the uneven heating of Earth sets up atmospheric currents that modify what would otherwise be extreme temperature differences between the equator and the poles. To this day, meteorologists call this current of air the "Hadley circulation," after the Englishman who thought of the idea in 1735.

A simple household analogy to this global current of air can be seen in a pot of boiling water on a stove. The stove supplies heat to the bottom of the pot, and the water loses heat by releasing steam at the top. Rising water at the pot's center carries the heat upward, and the returning, cooler (but still hot!) water sinks down near the edge of the pot.

This pattern of uneven heating and circulating currents of air changes during the course of a year due to the effect of the seasons. During the summer, the sun passes overhead at 23 degrees north latitude, while the overhead passage of the winter sun is at 23 degrees south. The latitude of the greatest heating of the ground is correspondingly farther north during the summer months, and the resulting atmospheric currents in turn shift to the north. Meanwhile, the North Pole, which sees no sunlight at all from October through March, is bathed in 24-hour sunshine during summer. This decreases the difference in net heating between the tropics and the Arctic, thereby weakening the heat-driven currents. Thus, seasonal variations in the solar heating of Earth's surface lead to shifts in weather patterns, with the patterns being, in general, farther north and weaker during summer.

AS THE EARTH TURNS

So now we have a simple heat-driven current that carries cold air to the equator and warm air aloft to the poles. That means that here in the West, the winds (at ground level) must always blow from the north, right? Of course they don't, and the reason is that the wind flow is complicated by the well-known fact that the Earth turns. The 7,918-mile-diameter Earth rotates once every 24 hours, carrying the people, houses, hills, trees and air on the equator around and around in an eastward direction at a zippy 1,038 miles per hour. Meanwhile, on the poles, where there are houses (sort of) and occasional people, things are carried along at a sluggish zero miles per hour. As the rising equatorial air moves toward the poles, its eastward motion of 1,038 miles per hour becomes increasingly greater than the eastward motion of the ground below. Unless something slows it down, the air will be moving 60 miles per hour faster than the ground at 20 degrees latitude, and 140 miles per hour faster at 30 degrees latitude. This rapid eastward flow of air is known as the subtropical jet stream, or just the "subtropical jet." It is one of the strongest and most consistent winds of the world.

By the time air reaches the subtropical jet, its eastward motion is far greater than the poleward motion, and our simple heat-driven circulation no longer is so simple. While the simple pattern can be seen in the tropics, up to about 30 degrees latitude on either side of the equator, it ends at the subtropical jets in each hemisphere. Farther from the equator, winds still blow east, but with nowhere near the regularity of the subtropical jet. This belt of winds is known as the "prevailing westerlies," from the meteorologist's (and mariner's) habit of referring to winds by the direction they blow *from*.

Meanwhile, up around the poles the simple circulation pattern tries to re-establish itself as radiatively cooled air piles up and heads south (in the Northern Hemisphere), but this, too, is broken up by Earth's rotation. In between, in that region known to meteorologists as the mid-latitudes, the poleward transport of warm air and the equator-ward flow of cold air is accomplished by a real mish-mash of swirling eddies of many sizes and shapes, all embedded in the general west-to-east flow of the prevailing westerlies. We know these eddies as cyclones and fronts, and they will be the subject of the next chapter.

WATER

Think of it—a world without clouds, where never a drop of rain or a flake of snow ever fell from the sky. Never a cooling fog or a dewy summer dawn, and no frost on the window. Pretty dull, eh? That is what weather would be like if there were no water vapor in the air. Remember the composition of the atmosphere, with its 99-percent nitrogen and oxygen and 1-percent other? Only a fraction of that remaining 1 percent is water vapor. And yet, slight as it is, that small concentration of water in the air is responsible for much of what we call weather. In fact, nearly all of the visible phenomena that make the weather so interesting are comprised of water in one form or another.

The importance of water in the atmosphere goes far beyond its role in making weather visually appealing. Water's place in the history of the Earth is assured, having kept our planet from becoming a Venus-like carbon dioxide pressure cooker. Even now, this most mundane of liquids is an essential ingredient of our weather. In its vapor state, water contains vast amounts of energy, which, when released, can literally unleash a storm. The energy in water vapor is called "latent heat." *Latent* means "present, but not visible or felt," and while latent heat cannot be felt as what we normally sense as heat (called "sensible heat"), it is very much present in moist air. Latent heat gets into water when it evaporates from the sea surface or wet ground. You know the cooling sensation of water evaporating from your skin—a cooling caused by water taking away your body heat. Where does this heat energy go? It goes into the water molecules and keeps them away from others of their kind, allowing the molecules to remain a vapor. When the vapor does condense back into liquid water, the latent heat is released into the atmosphere as sensible heat, where it causes rising currents of air.

Earlier, we talked about the heating of Earth by sunlight. Since 71 percent of Earth's surface is covered by water, most sunlight falls onto oceans. Earth's oceans weigh 1.5 billion billion tons, or 300 times more than the atmosphere. Most sunlight that strikes the oceans goes into heating that huge bulk of water, and very little goes into directly heating the atmosphere above. Rather, the atmosphere receives solar energy in the form of latent heat, as tropical breezes evaporate water from the sea surface. The amazing thing about latent heat is that it can be carried thousands of miles from its source before being let loose. Thus, energy absorbed from the sun by water vapor over the tropical Pacific Ocean may eventually fuel a storm over Montana. The mobility of latent heat allows the heat-driven motions of the atmosphere to concentrate over certain parts of the world, leading to some of the particular weather patterns that circle Earth.

GEOGRAPHY

Even if Earth were as smooth as a billiard ball, with no oceans, continents or mountain ranges, it would still have many of the familiar features of our weather, such as storms, fronts, jet streams and the like. But there would be none of the local variations that are so characteristic of western weather. San Francisco and Saint Louis, being at nearly the same latitude, could expect the same climate. This, of course, is not the case, and the reason is geography.

We have seen how the heating of Earth's surface by sunlight is the ultimate energy source that drives the weather. The way the sun heats the atmosphere is, however, profoundly affected by the nature of the underlying surface, be it lowland, highland or ocean. Furthermore, heat can be moved thousands of miles by ocean currents before entering the atmosphere as latent heat, and yet more thousands of miles by the atmosphere itself before being released as sensible heat. And mountains, by their sheer obstructive mass, can divert or destroy storms and shift jet streams.

The effects of geography on weather are as varied and detailed as geography itself. The role of geography in specific weather phenomena of the West will be pointed out in later chapters. But western weather is also influenced by one of the greatest geographical features on Earth—the continents and oceans of the tropics. Strangely, while most solar energy received by the tropics is absorbed by the oceans, the bulk of this energy ends up being released over land. Water vapor evaporated from the three tropical oceans—Indian, Pacific and Atlantic—is carried by trade winds to the three land masses of the tropics—Africa, South America and the islands of Indonesia. Over these land masses, latent heat is released in huge clusters of thunderstorms, resulting in rising currents of air that lead to powerful but localized Hadley circulation patterns.

The Hadley circulations over the three land masses are stronger than over the oceans and consequently generate subtropical jet streams that are faster and farther north. These speedy subtropical jets—with winds averaging 150 miles per hour—usually cross the Middle East, Japan and the southeastern United States. Elsewhere, such as over western North America, the subtropical jet is weaker and closer to the equator (and therefore well south of the western states). The winds of the subtropical jet often supply energy to mid-latitude storms, so we can begin to see how thunderstorms over Indonesia and South America can shape the weather of the West.

FRONTS, JETS, CYCLONES

In the previous chapter, we saw how the great winds of the world—the subtropical jet and the prevailing westerlies—are powered by sunlight falling on a rotating earth. Important as these currents of air are, they are not what we normally perceive as "weather." We think of weather as something smaller and more changeable, and indeed, less global and more "down home." Here in the mid-latitudes—including the West—it is the day-to-day passage of fronts and cyclones, of high and low pressure, that really makes up the weather. While these weather systems are not on as grand a scale as their globe-circling counterparts, they are caused by the same forces and are just as essential a part of the atmosphere.

To geographers, the mid-latitudes are the zones (one in each hemisphere) between 23 and 66 degrees latitude that separate the tropics from the Arctic and the Antarctic. Meteorologists define the mid-latitudes as the region dominated by the prevailing westerlies. Their location on the globe changes from day to day and season to season but, roughly, the mid-latitudes extend from 30 to 60 degrees latitude. More important, the mid-latitudes are where fronts and cyclones perform the vital task of carrying surplus heat northward from the tropics and returning cold air to the south. In the process, the opposing air masses clash and eventually mix, giving us our ever-changing weather.

THE MAKINGS OF A STORM

What happens when warm air meets cold air? First of all, there is a boundary between the two types of air, called a *front*. Yes, this is a term taken straight out of military books, by Norwegian meteorologists who came up with the concept shortly after World War I. Indeed, to this day fronts are shown on weather maps in a fashion identical to battle lines on military maps. Usually more than one front separates the tropical and arctic air masses. One front may divide tropical air from somewhat cooler air to the north, with another front between that cooler air and some downright nippy air north of that, and yet another between the nippy air and the bitterly cold arctic air near the North Pole. To understand these meteorological battles, we will first consider a single front.

Fronts are three-dimensional. There's cold air to the north, warm air to the south and a boundary between the opposing air masses that often runs east-west. Air masses also have depth, so fronts extend upward into the atmosphere. The upper section of the front usually looks quite different from the part near the ground. Meteorologists like to slice the atmosphere into five or six layers, each at a different height, to get a more complete idea of its workings. Let it suffice here to slice it into two layers—an upper level at 30,000 feet or so, and a lower level.

Pressure—that number you read off a barometer—is simply the weight of the overlying air. That's why the pressure goes down when you (and your barometer) go up. There's less air overhead at higher elevations. Furthermore, at the same pressure (and, roughly speaking, the same elevation), cold air is denser than warm air, meaning that the same amount of air is packed into a smaller volume. If you go up, say, 1,000 feet, you'll rise past more air if it's cold. Thus, in cold air the pressure drops faster with increasing altitude. Conversely, in warm air the pressure drops more slowly. The end result is that high above a front (say, at 30,000 feet), the pressure will be lower on the cold side of the front and higher on the warm side. Sometimes this is complicated by pressure differences across the front at ground level. If the surface pressure is higher on the cold side of the front, the pressure at 30,000 feet will be correspondingly higher on the cold side. Most of the time, though, this effect isn't great enough to change the final result: that cold air has lower pressure at high altitudes.

Pressure differences are important, since air tends to move from high pressure to low pressure.

Open the valve on an inflated tire, where the pressure is higher inside than outside, and you'll see what I mean. At 30,000 feet, for instance, air tries to blow from the warm side to the cold side of the front. But remember the earth is still turning. The warm air at, say, 40 degrees north and moving eastward with the earth at 796 miles per hour heads north to 50 degrees latitude, where the ground below is moving at only 668 miles per hour. The air ends up moving east at 128 miles per hour. Strong westerly winds (blowing *from* the west) are usually found along fronts at the upper levels of the atmosphere and are called jet streams, or just jets.

Unlike the subtropical jet, these frontal jets come and go as their fronts develop, move and dissipate. Frontal jets can sometimes form near enough to the subtropical jet to be able to pick up some of the subtropical jet's wind speed. Because the subtropical jet is usually fairly weak and well to the south of the western states, the frontal jet streams over the West tend to be weaker (but not always!) than they are over the East and elsewhere.

With cold air to the north and warm air to the south, a front extends in an east-west direction with a westerly jet blowing along its upper edge. This situation doesn't last long, since eventually the cold air starts moving southward, and the warm air northward. This *must* happen for the arctic and tropical air masses to mix. At this point, the east-west front begins to twist. The southward bulge ahead of the advancing cold air is called a "cold front," and, not surprisingly, the northward bulge leading the warm air is a "warm front." These are the beginnings of a storm.

Let's get back to the turning earth. We've seen, in several instances, northbound winds turned to the east by the rotation of the earth. Conversely, southbound winds are deflected west. In either event, the air flow is swerving to its *right*. No matter which way the air is moving, it always tends to curve right in the Northern Hemisphere

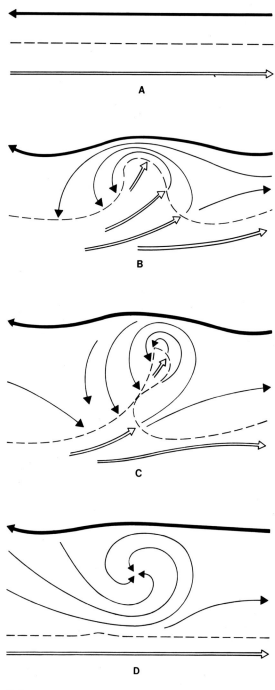

A

B

C

D

LIFE OF A STORM—a. The storm begins along the boundary between cold air to the north and warm air to the south. b. A ripple forms along the front, with cold air moving southward and warm air northward. c. The storm intensifies as the cold and warm air masses swirl around the center. Eventually, the faster-moving cold front catches up to the warm front, creating an occluded front. d. The fronts disappear when the air masses mix and blend together; all that remains is a dissipating swirl embedded in the original cold air mass.

(we see the opposite effect south of the equator). This is called the "Coriolis effect," after Gaspard Coriolis, the French engineer who, in 1843, proved it mathematically.

There are some important sidelights to the Coriolis effect. Initially, a low-pressure center sucks in air from all sides, leading to inward currents converging on the center. By bending each of these currents to its right, the Coriolis effect sends the air flowing *around* the low-pressure center in a counterclockwise direction (in the Northern Hemisphere). As always, the opposite holds true for high pressure. Now think about it. The counterclockwise winds blowing around a low-pressure center are also worked on by the Coriolis effect, which tries to make the winds start moving *away* from the low-pressure center. Meanwhile, the low pressure tries to suck the air back in. The end result is a sort of balance, with the air going around and around in a counterclockwise direction, but never heading into or away from the low-pressure center. In reality, this balance never exactly occurs, and some air does leak into or away from the "low," as meteorologists are fond of calling low-pressure areas. However, the near-balance keeps the low from either collapsing in on itself or flying apart and allows the storm to keep on spinning as long as it does (a week or so).

We return now to the bulging fronts. Take

the case of the cold bulge occurring west of the warm bulge. The cold air heads south and tries to turn west while the northbound warm air tries to head east. The two flows are trying to move away from each other. With air moving away in two directions, a partial vacuum develops in the middle. In other words, a low forms. The low pressure then counteracts the diverging warm and cold air flows, starts to pull them back toward the center, and we end up with a low center with cold air pouring down its western side and warm air streaming up its eastern side. This is called a "cyclone."

West of the cold bulge there may be another warm bulge. Between the bulges, the turning effects are the opposite of those leading to a cyclone, and the result is known as an "anticyclone." This is a region of high pressure, so it is called a "high." In the Northern Hemisphere, winds blow clockwise around a high. The eastern edge of the high, where winds are from the north, is colder than the western side.

Meanwhile, back at 30,000 feet, the jet is twisting with the bulging fronts. The jet dips south with the cold front and bows north with the warm front, and the line depicting the jet on a weather map may take on the sinuous appearance of a snake. The northward bends of the jet overlying the warm air are called "ridges." The southward-bending "troughs" are above the cold air, so pressure in the trough is lower than in the ridge. Near the surface, the cyclone remains in the middle of the whole mess, with the ridge to its east and the trough to its west. So a peculiar thing happens—the center of low pressure aloft is not above the surface cyclone, but to the west of it.

As the storm continues to develop, the cold air plunges even farther south. The trough often becomes so sharp that it breaks off from the rest of the jet stream, forming an "upper-level low." Sometimes the "upper low" (an interesting combination of terms!) separates from the low-level

storm, and the two go on their separate ways. This actually happens frequently in the West, where mountains have a way of destroying the low-level cyclone. When storms split like this, they often wither and die. Sometimes, however, the upper low may wander along and overtake another front, where it finds new life as another cyclone.

WEATHER PATTERNS OF THE WEST

Although jets may wiggle, their winds still blow, for the most part, to the east. Everything associated with a jet, including cyclones, fronts and even the wiggles, is carried eastward with the jet's winds. We may therefore think of jet streams as "storm tracks," the paths taken by cyclones as they live out their lives. This is one reason jet streams are such important features on weather maps—their average positions give a good idea of where the main storm tracks are.

Except for its Pacific Northwest coast, the West is the driest part of the United States. There are, of course, reasons for this. Lack of an available supply of moisture isn't the answer, because the whole Pacific Ocean lies right next door. Moist air doesn't do any good without some mechanism to wring water out of the air and drop it to the ground. Storms, which provide the most effective way to squeeze out this moisture, tend to skirt the western states. That is why the West is dry. Although it may sometimes seem otherwise, the West usually is ignored by the major storm tracks of the world.

Consider what causes storms. Sharply contrasting air masses are a must for healthy cyclones, and having a strong subtropical jet nearby doesn't hurt. In winter these conditions are met to a T over the eastern coasts of North America and Asia, where chilly air over the continents clashes with balmy air above the warm offshore currents. The West never sees the eastern North American storms, which track off in the opposite direction. Meanwhile, East Asian storms have to cross the

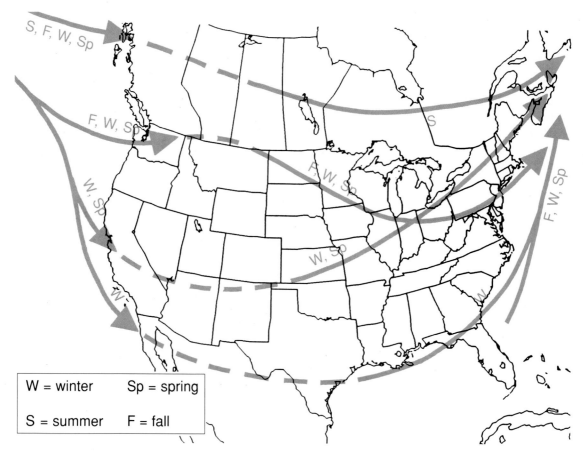

THE MAJOR STORM TRACKS OF NORTH AMERICA—Solid lines show the main storm tracks. Dashed lines show the frequent paths of upper-level storm remnants across the western mountains. The seasons when the tracks are most active are marked.

Pacific before reaching western shores. The crossing takes about a week, and by then many cyclones have lost much of their punch. Often, just the remains of the upper low make it to California.

Sometimes, however, storms are still going strong when they reach the West Coast, particularly if they get a boost from the temperature contrasts along the south coast of Alaska. But even a powerful storm striking the West Coast meets its match when it tries to cross the highlands of the West. The vast region between the Sierras and the Rockies is, on the average, the most elevated terrain in North America, with nearly a million square miles of land lying one to two miles above the sea.

Crossing a two-mile-high and 1,000-mile-wide mass of mountains is a real hurdle in the life of a cyclone. Think of the storm as a rotating swirl of air extending from sea level to six miles up. To cross the barrier, six vertical miles of storm must squeeze into just four miles. Like a stepped-on plum, the cyclone responds to this squeeze by spreading horizontally. This horizontal spreading slows the cyclone's rotation, just as a twirling figure skater who extends her arms will spin slower. This squashing all but destroys many storms and leaves others in a severely weakened state. We can see why many storms skirt along the Mexican and Canadian borders of the western states, where the mountains aren't quite as imposing. Many less-ambitious cyclones are content to remain just off-shore, spinning out their lives without ever seeing land.

If a storm squashes and weakens when it encounters high terrain, what does it do when it moves back over lower ground? As is so often the case in meteorology, the exact opposite happens. The storm stretches vertically, tightens inward horizontally and spins faster once again. Although the mountains weed out weaker storms, those storms that make it across stand a good chance of redeveloping when they reach the High Plains east of the Rockies. This is especially true if there is a strong frontal boundary crossing the plains, a common situation in the winter. The front gives a new source of energy to the storm system, and a new and powerful cyclone may sweep into the midwestern states. These ensuing storms have been named "Colorado" and "Alberta" lows, after the favored places for the regeneration process.

These are the storm tracks of winter. Pacific storms crash into the West Coast, especially in the Pacific Northwest, and break up as they try to cross the mountains. Their upper-level remnants track along the northern and southern borders of the United States and give birth to new storms when they once again reach lower ground and fresh fronts of warm and cold air. This double storm track is a peculiar feature of the West and leads to what is known as a "split jet." The powerful jet that frequently streaks across the North Pacific divides in two as it traverses the West, rejoining into one jet over the Midwest. Occasionally, the subtropical

jet joins the action, breeding upper lows south of Hawaii that sweep into the Southwest and southern Rockies.

During summer, the South American thunderstorms that kick off the Hadley circulation and the subtropical jet move into Mexico and Central America, while Indonesian thunderstorms migrate north to Southeast Asia. This pushes the Pacific and northern United States storm tracks farther north into Canada, and the southern storm track—the one that runs along the Mexican border in winter—disappears completely. Few major storm tracks cross the West in summer, and the area remains under an upper-level ridge. Sunny, dry weather is the rule; only infrequently do fronts and weak cyclones invade the West.

The cloudless skies of summer help per-petuate the upper-level ridge that caps the West at this time. Intense heating of the one- or two-mile-high ground by the nearly overhead sun warms the air at these altitudes. Warmer air leads to higher pressure aloft, and so the upper ridge, once it sets up in the early summer (usually in early June), becomes a persistent feature for the next three months.

Late in the season, the mass of tropical thunderstorms over Mexico may extend into the Southwest. This is the "summer monsoon," a brief rainy season that brings fleeting, but occasionally drenching, rains to the deserts. The ridge itself starts breaking down in late August or early Sep-tember, when the sun has dropped too low in the sky to provide enough heating to reinforce the ridge.

Storm tracks of the in-between seasons, autumn and spring, have features of both the summer and winter patterns. In general, the storm tracks are slightly farther north than in winter. The Canadian border track is well into Canada, while the Mexican border track cuts across the south-western states. The storms that take these paths are usually not as strong or as frequent as in winter. In spring, though, the regenerating storms along the east slopes of the Rockies can become quite intense. The boundary between lingering arctic air over the northern plains and moist air surging north from the Gulf of Mexico can breed powerful cyclones that bring heavy snows and rains to the eastern Rockies and High Plains, and tornadoes to the Midwest.

STORMY WEATHER

The passage of cyclones and anticyclones brings the West its never-ending variety of weather. Like snowflakes, no two storms are identical. And as cyclones progress through their life cycles, the type of weather they bring also changes.

However, most cyclones have some common features that are worth describing. The most important and simple guide to understanding the weather of high- and low-pressure areas is to remember that rising air brings clouds, rain and snow, while sinking air brings clear skies. Three principles of physics make this so: atmospheric pressure decreases with increasing altitude, air expands and cools when it rises into the realm of lower pressure, and cool air is less able to hold water vapor. So the vapor condenses back into droplets of liquid water (or crystals of solid water), forming clouds. With enough moisture in the air, the droplets and crystals coalesce into raindrops and snowflakes.

AUTUMN COLD FRONT—As it plunges southward, the heavier cold air behind the front flows like molasses spilled onto a table—only faster. The leading edge of the cold air—the cold front—takes on the same curved shape as the front edge of the spreading molasses. Moisture in the cold air mass makes this front visible as a fog bank advancing into the Colorado Rockies.

Richard A. Keen

AERIAL VIEW OF DYING PACIFIC STORM—This dying storm over the North Pacific Ocean was photographed from earth orbit by the Apollo 9 astronauts. At first glance, the storm looks like a lively one, with a well-developed spiral flow into its center. However, the clouds are small and broken, and a ship passing through this storm would encounter nothing more than light rain or drizzle.

NASA

The twisting fronts around a cyclone determine where the air currents rise and where they fall. Since most lows move east, the northward-moving warm front is ahead of the low center—the "warm before the storm"—and the southbound cold front follows the passage of the low. Since warm air is lighter, it tends to overrun the cold air. This usually brings thickening and lowering clouds followed by steady precipitation in the form of rain or snow. As it plunges south and east, the heavier cold air cuts under the lighter warm air. The undercutting cold front shoves the warm air upward, often in a fairly narrow line, setting off brief but heavier precipitation in showers, squalls and thunderstorms.

Many cyclones reach the West Coast near the end of their lives. These old storms usually don't have well-defined fronts and can be best described as spinning whirls with little up-and-down air motion. The weather is as lackluster as the cyclones themselves, often appearing as drizzle along the coast and light snow in the mountains. Quite a few of these old storms drift slowly southward along the coast, finally retiring off the shores of Mexico. Others move inland, occasionally finding new life as regenerating storms.

There is nothing like a good mountain range to complicate this picture of the sequence of weather brought by a passing cyclone. Simply put, when winds reach a mountain range, they blow up and over the ridge and down the other side. Since rising air brings precipitation, the *upwind* side of a mountain range gets an extra dose of rain or snow. With prevailing westerlies most of the year, the western slopes of the Cascades, Sierras and Rockies get quite a boost to their annual precipitation from upslope winds. Sometimes in these areas, rain or snow continues for a day or more after the skies have cleared elsewhere.

The moisture from rain and snow that soak the western slopes does not fall somewhere else. That somewhere else is the eastern slopes of the western ranges. Prevailing downslope winds give many of these regions a desertlike climate; all of Nevada and parts of eastern Washington, Oregon and California are clear examples of this so-called "rain shadow" effect.

The eastern slopes of the Rockies are also subject to rain shadowing much of the year. However, springtime in the Rockies is a different story. Regenerating cyclones over the High Plains draw humid air from the Gulf of Mexico and swirl it around the north side of the low center, right into the Rockies. Riding winds from the east, this Gulf moisture can be unloaded in copious quantities onto the foothills. The world-record heaviest snowfall, seven feet in just over a day, was dumped on Colorado during one of these springtime reversals of the prevailing westerlies.

The south-facing mountains of northern Arizona, New Mexico and southwestern Colorado see upslope precipitation when the winds are from the south, often ahead of an approaching cyclone. With the convoluted terrain of the West, it is easy to realize why the weather can vary so greatly from one county to the next.

The cyclone passes, and its potpourri of precipitation has come and gone. It is now the anticyclone's turn to take charge of the weather. This high brings clearing and cooling, but, as the high moves east, southerly winds usher in warming. High-level clouds appear on the horizon, foretelling the approach of the next cyclone.

At many places in the mid-latitudes, cyclones pass at three- or four-day intervals. This rule of thumb works well in the eastern states. But the West has its "split jet" and double storm track. Frequently, alternating cyclones take each route: one goes north, the next goes south. On each storm track, cyclones pass once a week rather than every three days. What this means for westerners is that storms often arrive at weekly intervals. But sometimes the jet doesn't split or only splits after it has reached well inland. These weather patterns

Richard A. Keen

GYPSUM STORM—A crowd awaiting the landing of the space shuttle Columbia *at White Sands, New Mexico, watches apprehensively as high winds whip clouds of gypsum thousands of feet into the sky. Poor visibility on the landing strip forced a one-day delay in the landing.*

can subject parts of the West Coast to a near-continuous battering by Pacific storms, with the skies scarcely clearing after one storm before clouds from the next cyclone start rolling in. There are also the occasional "blocking" weather patterns, so named by the action of large, slow-moving highs or lows (or a combination of both) that block the regular west-to-east flow. These sluggish weather patterns can bring days of lingering inclement weather followed, fortunately, by an equally long spell of fine weather.

One reason western weather never gets boring is that weather patterns can change several times during the course of a season. Months may

pass without a good, healthy storm, but then the pattern breaks. Suddenly, cyclones queue up across the Pacific from California to Japan, each impatiently waiting its turn to march ashore. Then, unexpectedly, they're gone. Watch the weather maps and satellite photos in your newspapers and television broadcasts and you'll see this happen. It's fascinating!

MEMORABLE WESTERN CYCLONES

November 1861–January 1862—A non-stop series of Pacific cyclones struck the entire West Coast, causing severe flooding from San Diego to Olympia, Washington. The central val-

21

leys of California turned into a "sea," and, up in Olympia, a resident wrote that "everybody's roof leaked."

January 1921—The "great Olympic blowdown" swept the coasts of Washington and Oregon with gusts as high as 150 m.p.h. Gales destroyed eight times as much Douglas fir lumber as did the explosion of Mount Saint Helens in 1980.

October 1934—Hurricane-force winds raked western Washington, unroofing buildings and sinking boats in Puget Sound. One casualty was the liner *President Madison*, which ripped loose from her moorings, hit several other ships and sank.

November 1940—Not a really big storm, but enough to send the Tacoma Narrows Bridge in Washington into the drink. Flaws in the bridge's design caused it to vibrate and twist when winds were just right. Already bearing the name "Galloping Gertie," the bridge was unable to survive this storm. One dog reportedly died in the collapse.

October 1962—The "Columbus Day storm" in western Washington and Oregon was, in part, the remnants of Pacific typhoon "Freda." It was possibly the most powerful storm ever to pound the West, with winds reaching 170 m.p.h.

January 1969—Ten days of storms brought floods, mudslides and falling trees to much of California. More than 40 people drowned or were buried in mudslides.

March 29, 1982—Gusts of nearly 60 m.p.h. lifted huge clouds of gypsum dust above the landing strip at White Sands, New Mexico. The near-zero visibility forced the space shuttle *Columbia* to delay its landing until the next day.

November 1982–April 1983—One of the strongest and most extended series of cyclones in memory pounded the California coast. Storms brought wind, rain, surf and even tornadoes to southern and central California, damaging or destroying more than 3,000 homes and businesses. Many of the storms continued into the Rockies with heavy snows—a Christmas blizzard shut down Denver with two to four feet of snow, and an April storm dumped 65 inches near Fort Collins, Colorado.

February 1986—Twelve days of torrential rains, totaling as much as 49.6 inches, drenched northern California. Dam and levee breaks flooded cities in the Sacramento Valley, including Yuba City and Linda, causing nearly half a billion dollars of property damage.

WATER FOR THE WEST

One of the most remarkable features of the climate of the West is its fantastic variability from place to place. In many parts of the West an easy hour's drive takes you from verdant forest to sagebrush desert. To a large extent, these stark contrasts of vegetation are due to differences in the amount of moisture that drops from the sky. This moisture falls in many forms, some liquid and some frozen. Rain, sleet, hail, snow—all those inclemencies that mail carriers must go through—are all lumped together under the common name "precipitation."

Precipitation is the depth of rain, melted snow, hail and other frozen stuff that would be standing on the ground after a storm if none ran off, soaked into the ground or evaporated. The measurement of precipitation reflects this concept. Frenchman Denys Papin put out a bucket to measure the rain in 1669. And ever since, no matter how sophisticated they may look, most measuring devices for precipitation are still essen-

tially buckets. Some use heating elements or automobile antifreeze to melt snowfall and some have digital readouts. But all are still cans that show the water depth at the end of a storm. Add these measurements over the course of a year and you get the annual precipitation for your locale. Do this for several years and you can get the average annual precipitation. This number—average annual precipitation—is one of the single numbers that best describe the climate.

The wettest weather station in the West is Wynoochee Oxbow, Washington, on the Olympic Peninsula, with an average annual rainfall of 144.43 inches. That is 86 times as much as the West's driest spot, Death Valley, California, with only 1.68 inches a year. The extreme variability of the West contrasts with the evenness of annual rainfall across the eastern states. For the 26 states east of the Mississippi River, the extremes are 80 inches in the hills of South Carolina and 28 inches in western Wisconsin. Amazingly, this range of

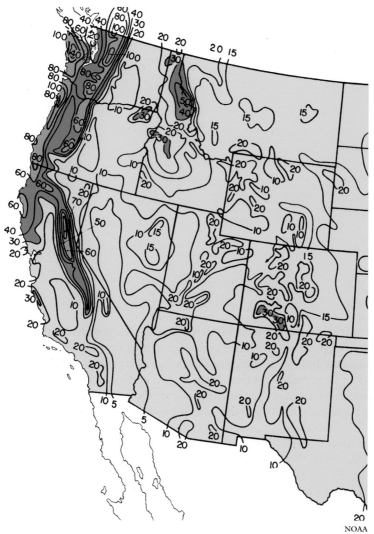

AVERAGE ANNUAL PRECIPITATION IN THE WEST, 1899–1938

NOAA

average annual precipitation can be found between sites 12 miles apart within Clallam County, Washington, and 16 miles apart in Shasta County, California.

As best as it can be figured, the area-averaged annual precipitation for the 11 western states is about 17 inches. The rest of the "lower 48" states average over twice as much—about 36 inches. While the West may be the driest part of the country, the amount of water that falls is nonetheless enormous. Seventeen inches a year comes out to half a million gallons per acre. Over the 1,193,826 square miles of western states, the annual soaking amounts to 319 *cubic* miles of water, or 1.5 million million tons of the stuff. That's 10 times the volume of Washington's

Puget Sound. The amounts are awesome. Where does it all come from?

Water vapor is always part of the atmosphere overlying the western states. The amount of this vapor varies considerably from season to season and even from day to day, but on the average the total volume of moisture over the West is about 10 cubic miles of liquid equivalent. This is enough to cover the region with a half inch of liquid water. However, again on the average, a half inch of precipitation falls on the West every 10 days. This means there must be a continuous replenishment of atmospheric moisture over the West.

There is no single source for the West's water. Sources change from state to state and from season to season. We know that moisture evaporates off the surfaces of the world's oceans because, ultimately, that is where it all returns. In fact, oceans are the West's primary water source. The cycle of evaporation from oceans, transport by winds to land areas, precipitation, runoff into rivers, and return to the oceans is known as the "hydrologic cycle."

In reality, the hydrologic cycle is complicated. Water can be stored for long amounts of time in lakes, soil, underground water supplies and even in snowpack and glaciers before being released back to the oceans. Even atmospheric transport of moisture can be circuitous—water can precipitate and evaporate several times before reaching its "final" destination.

WINTER MOISTURE

Maps of precipitation distribution across the West, and of the seasons in which most of it falls, provide strong clues as to its source. In general, the west coast states are wetter than the interior states and, in these coastal states, the western slopes of the mountains are wetter than the eastern slopes. Most of the annual precipitation in these regions falls during the winter, when the prevailing westerlies are most pronounced. It

is therefore a simple deduction that the West Coast states, and the western slopes of the interior mountain ranges, get most of their moisture from the Pacific Ocean.

It is no big surprise that the Pacific supplies much of the West's moisture, considering its size and placement. However, the places that rely on Pacific moisture are the same places that are dry during summer. The ocean is still there, of course, but the means of getting Pacific moisture onshore

RAIN (OR SNOW) SEASONS ACROSS THE WEST—In some parts of the West, winter is the wettest season of the year, while other regions receive their moisture during summer. This map shows which of the four seasons receives the most precipitation for different areas.

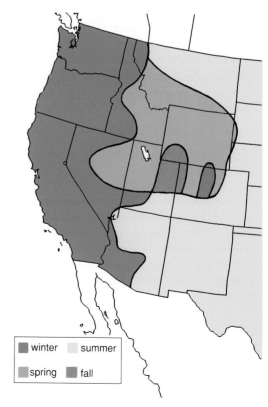

winter   summer

spring   fall

fall
winter
spring

summer
fall

spring

summer

THE MAIN STREAMS OF MOISTURE INTO THE WEST—In a schematic way, the arrows depict the flow of water vapor from the North Pacific Ocean, the Gulf of California and nearby areas of the Pacific, and the Gulf of Mexico, along with seasons each moisture stream is important. The width of the arrows gives a rough idea of the relative importance of each of the moisture sources.

and then squeezing it out of the air is not. The storms that lash the coast during the winter are gone, and only occasionally do weak storms brush the Pacific Northwest. Summer is dry on the West Coast. Mark Twain succinctly summarized the nature of West Coast rain in his description of San Francisco's weather in his book *Roughing It*:

> When you want to go visiting or attend church, or the theatre, you never look up at the clouds to see whether it is likely to rain or not—you look at the almanac. If it is Winter, it will rain—and if it is Summer, it won't rain, and you cannot help it.

SUMMER MOISTURE

Summer is the wettest season over much of the interior Southwest and along the High Plains

east of the Rockies. Each of these regions has its primary moisture source. The southwest region—Arizona, southern Utah, western New Mexico and southwestern Colorado—gets its moisture from the warm waters of the Mexican Pacific coast and the Gulf of California. These tepid waters can be considered separately from the main mass of the Pacific Ocean that supplies winter moisture to the coast, because each feeds moisture to a different region at a different time of year through the action of a different kind of storm. Winter moisture is brought by cyclonic storms, while most summer rains fall in thunderstorms. The summertime flow of tropical air into the desert Southwest is called the "Southwest monsoon," after its larger and stronger Asian counterpart.

Like its Asian cousin, the Southwest monsoon can vary from year to year and even during the course of an individual season. It is often intermittent, with rainy spells of several days or weeks duration interspersed with similarly long dry spells. Some years the monsoon barely reaches into southern Arizona, while other years its thunderstorms can be followed into Idaho and Oregon.

The other region with predominant summer rains, the High Plains, stretches 1,300 miles from border to border. This area includes western Texas, the eastern sections of New Mexico, Colorado, Wyoming and Montana, and the Black Hills of South Dakota. The imposing barrier of the Continental Divide separates this region from the tropical moisture of the Southwest monsoon. In its place, water vapor streams in from the Gulf of Mexico to feed the summer storms of the High Plains. At this time of year the Gulf of Mexico simmers at 85 degrees, allowing large amounts of water to evaporate into the overlying air. The Gulf is such a prodigious source of moisture that it supplies much of the rain that falls over the entire United States east of Rockies. During the summer, some of it finds its way toward the Rockies.

The river of humid air flowing north from

the Gulf of Mexico normally stays well east of the Rockies. The summertime boundary between the usually arid air of the West and the humid mass of Gulf air often lies near the 100th meridian, several hundred miles east of the Rockies. Frequently, this boundary forms a sharply defined front called the "dry line." The dry line separates wet and dry rather than warm and cold air masses, but it is every bit as important to the daily weather of the summer months. Severe thunderstorms, and sometimes tornadoes, like to form along the dry line; however, most of the time these storms form too far east to be considered western weather.

Like other fronts, the dry line does not stand still. Its westward excursions allow Gulf air to flow right into the foothills of the Rockies. This happens often enough to bring five to 10 inches of summer rain all along the High Plains.

The moist streams from the two Gulfs, the Gulf of California and the Gulf of Mexico, sometimes meet over the Rockies of northern New Mexico. The result is the most pronounced summer rainfall maximum in the West, with some spots receiving half their annual precipitation in July and August alone. Most of this rain falls in thundershowers, which occur more often here than anywhere else in the country.

SPRING MOISTURE

Never mind what the song says; springtime in the Rockies is the *wettest* time of year. Along the Rockies and their eastern foothills from Denver northward, the three spring months—March, April and May—see more rain and snow fall from the sky than any other season. In these areas summer is the second wettest season, and the source of the moisture is the same—the Gulf of Mexico. The dry line is there too, especially in late spring. However, the means for bringing moisture to the mountains is a lot more vigorous in the spring than it is in the summer.

The eastern plains of Colorado are a favored

location for the regeneration of cyclones that make it across the mountains. The counterclockwise rotation of these storms swirls the humid Gulf air north and then westward around the storm's center, flinging moisture against the slopes of the Rockies. Although this may happen only a few times a year, the resulting precipitation (often snow) can be extremely heavy.

AUTUMN MOISTURE

Autumn is the wettest season in just a small area of northeastern Utah and northwestern Colorado, and even here it is just barely moister than the other seasons. Located on the northern fringes of the Southwestern monsoon, this area's autumn maximum is earned in those occasional years when the monsoon continues into September. August, September and October have been known to bring hurricane remnants into the southwestern states and, although rare, their rains can be torrential. Over the long run, hurricanes provide a significant portion of the late summer and fall precipitation for much of the Southwest.

LOCAL MOISTURE SOURCES

There are three main oceanic sources of water for the West—the Pacific Ocean, the Gulf of California and the Gulf of Mexico. However, the path of a water molecule from the ocean to your backyard may not be a direct one. Much of the moisture blown in from the ocean and dumped on the West collects on the ground and in lakes, where it evaporates and becomes an important moisture supply for later storms. It is quite possible that some of the water in a local rainstorm has been "recycled" several times since it evaporated from the ocean surface.

One place water can be stored for recycling is in the soil. Heavy storms are often followed by lighter showers fed, in part, by moisture evaporated from the soaked ground. Storage of moisture in the soil is one way that wet spells and, conversely, droughts perpetuate themselves. In summer droughts, rain showers are less likely to develop as long as the ground remains dry. Dry ground is more readily heated by the summer sun, since less solar energy goes into evaporating moisture from the soil. The more efficient heating of dry ground and of the air above it helps maintain the upper-level ridge, which, in turn, helps suppress the development of showers.

Mountain snowpack can have the same effect as wet soil; the major difference is that the moisture is lying on top of the ground. Evaporation from melting snowpack is one source of water vapor for showers forming over mountain areas in late spring and early summer.

Perhaps the West's most unique local moisture source is its largest body of surface water, the Great Salt Lake. Lakes of this size deliver large

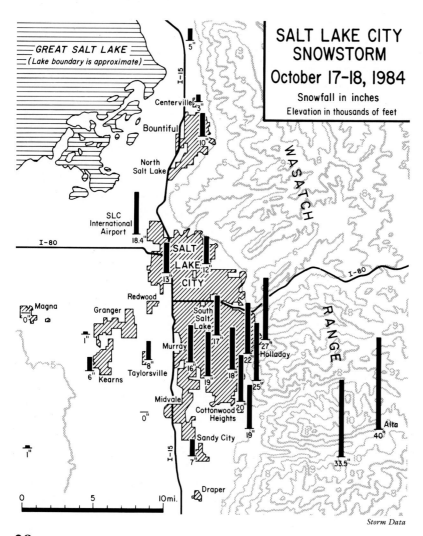

SALT LAKE CITY
SNOWSTORM
October 17-18, 1984
Snowfall in inches
Elevation in thousands of feet

GREAT SALT LAKE
(Lake boundary is approximate)

Storm Data

Richard A. Keen

A MANMADE CLOUD—When air is cool and moist enough, the added moisture from steam-cooled power plants can be enough to trigger formation of a cloud. The temperature was near zero when this photo was taken of the Public Service plant in Boulder, Colorado. Most of the moisture going into the cloud is coming from the heated water of the small lake to the right of the plant.

amounts of water vapor into the atmosphere, particularly when the lake is warmer than the air blowing across it. This occurs in late autumn and early winter, when the first outbreaks of arctic air reach a lake still retaining the warmth of summer. On the coldest days moisture can be seen entering the atmosphere as twirling streamers of mist above the lake surface. Take a cup of hot coffee outside on a winter day and you'll see this in miniature. Usually, this moisture doesn't rise very high into the atmosphere; but when the lake is warm enough and the air cold enough, low clouds and "lake-

effect" snow can develop over the downwind side of the lake.

One lake-effect storm dumped anywhere from zero to 40 inches of snow on different parts of the Salt Lake City area in October 1984. Between the lake and the crest of the Wasatch Mountains, some 50 million tons of snow dropped from the sky—enough to make a mile-high snowman!

How small can a lake be and still make a snowstorm? Little lakes, perhaps a mile or less across, usually freeze over when the air gets really

29

cold. But when a lake is used to cool the burners in factories and power plants, its water stays warm all winter long. Occasionally—when the atmospheric conditions are just right—miniature snowstorms coat the ground downwind from factory-heated lakes and even large cooling towers. One of these "backyard blizzards" whitened a two-mile-wide section of Boulder, Colorado, near the local power plant one January morning. There was only a skiff of snow, enough to fill the cracks in the sidewalks, but piled all in one place it may have weighed five or 10 tons. That's not a whole lot of snow, but these little storms do illustrate the hydrologic cycle on a scale that we can grasp.

The Great Salt Lake may be enormous compared to a heated pond, but it is still puny compared to the vast expanse of the Pacific Ocean. Likewise, its real importance as a moisture source is relatively unimportant. The big lake-effect snow in 1984 supplied only .003 percent of the West's total precipitation that year. We are left realizing the real sources of water for the West are the surrounding oceans, and that some of this moisture entered the atmosphere half a world away.

A BACKYARD BLIZZARD—Sometimes, snow may fall from clouds created by power plants. On this occasion, the snow from Boulder's power plant dusted a two-mile-wide area. Within the heart-shaped area there was enough snow to whiten the ground.

source: David Blanchard and Stan Barnes

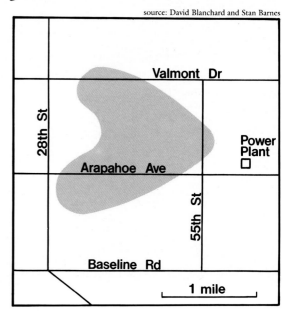

S N O W   A N D   R A I N

In the past three chapters we have seen how the sun makes the wind blow and how the wind brings enormous amounts of water from the world's oceans into the West. That leaves us with five or 10 cubic miles of water sitting over the western states on an average day. However, this water is in the air and, instead of being a refreshing liquid that animals and plants can enjoy, it's all vapor. That brings us to the question of how to get all this water out of the sky and onto (and into) the ground.

CONDENSATION

Water vapor is a gas and will never fall. To return to earth it needs to turn into liquid water or solid ice through a process called "condensation." Did you ever notice how your laundry on the line dries faster when it's warm outside? Warm air has less trouble evaporating water because it can hold more water vapor than cold air can. If air containing water vapor gets chilled, it eventually reaches

a point where it can no longer hold its supply of vapor. The excess vapor condenses into tiny drops of liquid water. If the air is cold enough, the vapor may form ice crystals through a process called "sublimation." Later we'll see that vapor can condense into liquid water even at temperatures well below the freezing point, although the drops may later freeze.

One way to describe the amount of vapor in the air is to measure the temperature at which condensation occurs. This temperature is called the "dew point," and the higher the dew point, the more moisture the air contains. Roughly, the amount of vapor in, say, a cubic foot of air doubles for every 20-degree increase of the dew point. An 80-degree dew point means four times the moisture as 40 degrees.

There's an old army saying that "the only things that fall from the sky are rain and paratroopers." Of course, meteorologically speaking, it just isn't true. It's not just rain and snow, either. Water can condense into an astounding variety of shapes and sizes before it falls from the sky. The form of precipitation that reaches the ground tells you a lot about what's happening miles overhead.

CLOUDS

Most of the time, there must be clouds before there's precipitation. Amazingly, if the atmosphere were completely "pure" and contained only gaseous molecules like oxygen and water vapor, we would never see a cloud. Even though air may cool below its dew point, water vapor still needs something to condense *on*. Water can't just condense in thin air. Fortunately, the air isn't pure; it is chock-full of minuscule particles of dust, pollen, sea salt, smoke and even tiny drops of turpentine blown from trees. The amount of these "condensation nuclei" varies greatly from place to place, but usually there are between a thousand and a million of them in every cubic *inch* of air. Of course, there are more above cities than over countryside and more over land than over sea. However, water vapor molecules rarely have to go very far to find a place to condense.

There are two main ways to cool air down to the dew point. The first is to lift the air; as it goes higher, the pressure around it falls off, and the air expands and cools. At a given altitude the temperature falls to the dew point, and condensation occurs. This altitude is the "condensation level"; above this level sits a cloud. Often, this lifting of air takes place in cyclones, as described earlier. Sometimes, particularly in summer, solar heating of the ground can send warm "bubbles" of air up to the condensation level. Another way to lift air is to simply blow it up a mountainside.

The second way to chill air is to set it out at night and let its heat radiate into space. At first, most of the radiation leaves the ground, not the air, and the air gets cold by contacting the cold ground. When stones and blades of grass on the ground cool below the dew point, dew or frost condenses; when the air gets to the dew point, a cloud—fog—forms near the ground. Blanketed by a layer of fog, the ground no longer radiates its heat directly to space. Now the radiational cooling occurs at the top of the fog bank, and the fog thickens and deepens. This is how valley fog forms. A variation on this theme is to run air over a cold surface, be it ground or ocean. Fog off the California coast develops this way.

The drops of water that condense to make clouds are incredibly small. They average much less than a thousandth of an inch in diameter, and it would take 10,000 of them to cover the head of a pin. Drops this small would take a week to fall to the ground, except that they never make it. Once outside the cloud, they evaporate in minutes. These lazily floating droplets are the answer to that nagging question of childhood: how do clouds stay up? But don't be fooled. There are a *lot* of these droplets in a cloud, and even the smallest puffy cloud on a summer afternoon can easily outweigh an automobile.

RAIN

We know that rain falls and clouds don't, so raindrops must be a lot heavier than cloud droplets. It takes about a million cloud droplets to make one average raindrop. Initially, a cloud droplet grows by condensing more water vapor onto its tiny spherical surface. Eventually it gets large enough to start falling through the cloud. On its downward journey, our growing drop bumps into and absorbs some of the many droplets still hanging in the cloud. This growth process is called "coalescence." When the drop gets as big as a sixteenth of an inch across, it's a genuine raindrop and down it goes.

Coalescence works well in thick clouds with a good supply of very moist air, so the droplets have plenty of opportunities to bump into each other. This happens a lot in the tropics. Coalescence doesn't work very well, however, in the cooler and drier climate of the West. If clouds are thinner and there's not as much water vapor around, drops don't grow to full size. These immature drops fall as drizzle, a common form of precipitation from the thin clouds and fog along the coast. If the air doesn't have the moisture to form even a drizzle, the cloud releases no precipitation.

SNOW

While coalescence may seem like a perfectly reasonable way to make rain, the truth is that in the West, and everywhere else outside the tropics, most rain is really melted snow (or hail, in the summer). This is partly because in mid-latitudes any cloud big enough to precipitate is probably tall enough to have its top portion well below freezing. Another factor is the way snowflakes grow faster than raindrops.

Raindrops may begin their lives as snowflakes, but there's yet another twist: snowflakes start out as water droplets! When water vapor condenses out of the atmosphere, it nearly always does so as droplets of *liquid* water. When the air is colder than 40 below zero, vapor may condense into minute ice crystals, but this is relatively rare. When the air temperature is between 32 above and 40 below, condensation results in peculiar creations: droplets of "supercooled" water whose temperature is below freezing.

You've probably heard stories about ponds in the north woods that have cooled below 32 degrees but haven't frozen over. An unfortunate duck alights in the water, and the pond immediately freezes around the poor critter's feet. There's reason to believe that ducks never suffer this ignominy, but the duck's tale does illustrate a point: supercooled water freezes immediately when disturbed. In the air, disturbances come as little particles called "freezing nuclei." Just as condensation nuclei start the condensation of vapor into droplets, freezing nuclei start the freezing of droplets into crystals.

Most freezing nuclei are the same kind of dust and smoke particles that work so well as condensation nuclei, although it may be that nuclei with different shapes and sizes do better at freezing. Even meteor dust—the ashes of "shooting stars"—might do the job. Then there are the artificial nuclei, such as silver iodide crystals, that cloud-seeders intentionally put into the air to trigger the freezing. Duck feet do not work well.

Once a droplet freezes into a little crystal, it turns into a wolf among sheep. Ice particles act as both condensation and freezing nuclei, grabbing moisture out of the air more efficiently than water droplets could. Growing ice crystals even steal moisture from nearby droplets and become snowflakes within 10 or 20 minutes.

Snowflakes are famous for their variety of six-pointed shapes, and it's probably true that no two are alike. Each snowflake is made up of thousands of billions of billions (otherwise known as sextillions) of water molecules that can be arranged in quadrillions of septillions of centillions of ways. The number of flakes that fall annually on earth is in the sextillions—far, far fewer than the number of possible different flakes. So it's likely that in the entire lifetime of our planet, there will never be a snowflake that's identical to any other past, present or future flake.

All these centillions of possible snowflakes can be lumped into several categories. The category a snowflake will fall into is pretty much decided by the temperature at which the flake forms. Flakes that grow from the skimpy water supply at 20 below or colder form six-sided (hexagonal) "columns" that look like pieces of a six-sided pencil. Around zero to 10 below the crystals take the shape of flat, six-sided wafers called "hexagonal plates." With the more plentiful supply of water vapor at zero to 20 above, crystals can grow into the large but delicate six-pointed stars that typify our idea of snowflakes. These crystals

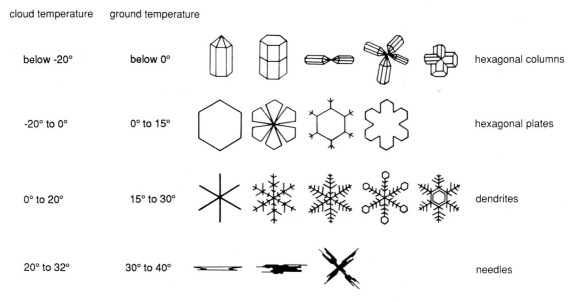

cloud temperature	ground temperature		
below -20°	below 0°		hexagonal columns
-20° to 0°	0° to 15°		hexagonal plates
0° to 20°	15° to 30°		dendrites
20° to 32°	30° to 40°		needles

from Nakaya

TYPES OF SNOW CRYSTALS

are called "dendrites," from the old Greek word meaning 'branched like a tree.' The warmest crystals, those that grow between 20 degrees and freezing, are splinter-shaped bits of ice known as "needles."

In moderate snowstorms, the individual crystals can reach the ground without hitting and sticking to other crystals. With a magnifying glass you can see an incredible variety of crystal shapes. They show up well on dark surfaces, such as wooden railings or winter coats. When snowfall increases, however, crystals stick together. These aggregates can get quite large at times, sometimes growing to several inches across.

For skiers, the best snow is the light powder, which means the aggregate flakes contain lots of air and little water. The biggest crystals—dendrites—make the fluffiest flakes. The biggest dendrites form when air temperature is about 5 above. Since most snowflakes in a snowstorm are formed

several thousand feet up, ground-level temperature where these flakes fall is 20 or 25 degrees. What this means is that the fluffiest, lightest snow falls when the temperature at ground level is in the low 20s. On the average, 13 inches of this snow melts down to an inch of water.

When the ground-level temperature is closer to freezing, snow comes down as bunches of needle crystals, looking somewhat like a handful of pine needles. These more densely packed crystals make a heavier layer of snow on the ground; 10 inches of snow makes only about an inch of water. Interestingly, when the ground-level temperature is below 10 degrees, the snowfall also gets denser. The little hexagonal plates and columns don't stick to each other very easily, and the layer they make on the ground doesn't hold very much air. However, it never gets too cold to snow—Yellowstone National Park has had snow with temperatures of 20 below or colder. Extremely cold air doesn't

34

contain much moisture, but what there is can be made into ice crystals. There just isn't very much of the stuff.

DIAMOND DUST

When it's *really* cold, zero or below, cloud droplets don't last too long before they freeze. This means extremely cold clouds are made largely of ice crystals, most of which are hexagonal columns and plates. When fog forms at extremely low temperatures, it, too, freezes quickly into crystals. This "ice fog" is common during Alaskan winters. Often the air is full of fairly large plate crystals falling slowly from a clear sky—there are so few water droplets that any fog or cloud is invisible. The twinkling and glistening of these crystals in sunlight has earned the spectacle the name "diamond dust." The meteorological name is more direct—"ice crystals." At night the crystals sparkle in flashlight and headlight beams, as thousands of flat plates reflect the light right back at you.

Diamond dust crystals fall with their flat sides parallel to the ground. This cloud of minuscule mirrors, each facing straight down, leads to one of the weirdest nighttime lighting effects the weather can produce. All along the horizon, above each and every distant street lamp and approaching headlight, are long, straight and narrow rays pointing straight up like searchlight beams—light reflecting off millions of crystals between you and the light source. The shifting beams above moving cars can even look like the northern lights!

DIAMOND DUST—These "searchlight beams" are really the reflections of street lamps off millions of tiny, flat ice crystals suspended in the air. These crystals often form in clear air when the temperature is below zero.

Dale P. Cruikshank

SNOW GRAINS

Sometimes, when a cloud is warm enough (but still below freezing), supercooled droplets can coalesce to the size of drizzle before freezing. If the cloud deck is fairly thin, perhaps less than a thousand feet from top to bottom, the frozen droplets leave the bottom of the cloud before they grow much more. The white, sand-sized bits of snow that fall to the ground are called "snow grains." Few roads have been closed by falls of snow grains, which rarely amount to much more than a dusting.

GRAUPEL

Yet another form snow can take is "snow pellets," also known as "soft hail" and, more commonly, by the German name, *graupel*. These roundish or cone-shaped, BB- to marble-sized snowballs usually come down in quick showers and bounce once or twice. Heavy graupel showers sometimes come with thunder and lightning. These snow pellets grow as turbulent air currents inside the cloud carry them through swarms of water droplets, which freeze onto their crystal structure. Since they pick up whole droplets, snow pellets are heavier, grainier and lack the delicate features of snowflakes.

GLAZE ICE

On a sultry summer afternoon, it just doesn't seem believable that the rain shower steaming from the sidewalks started out as a snow-storm. But the higher you go, the colder it is, and eventually the temperature falls below freezing. This "freezing level" changes from day to day, but even in the heat of summer—indeed, even in the tropics—it never gets higher than 19,000 feet above sea level. Summer thunderheads often tower to twice that height, so the top halves of their clouds actually harbor raging snowstorms. At a rate of 3 to 5 degrees every 1,000 feet, the snow-flakes warm as they fall. When they drop below the freezing level, they melt into raindrops. Most of the time the raindrops stay melted and soak into the ground or run off into streams. However, a nasty thing happens if there is a layer of cold air near the ground: the rain freezes on everything in sight. Trees, wires, grass and even roads can be coated with a sheet of glossy ice called "glaze" or, appropriately, "freezing rain."

The beauty of these "ice storms" can't be denied; even the most mundane of objects, from trees to television antennas, assume the elegance of

RIME UP CLOSE—This close-up photograph of rimed pine needles shows the true nature of rime. The rime spikes, less than a quarter of an inch long, are each composed of hundreds of frozen water droplets.

Richard A. Keen

fine crystal. Unfortunately, the reality is that glaze often reduces the value of whatever it touches. Heavy accumulations bring down wires and branches, and driving and walking can become treacherous. Glaze ice is usually a fraction of an inch thick, but two-inch-deep coatings have been seen on wires and trees. At that rate, a 100-foot length of wire would be burdened with 600 pounds of ice, and the branches of a typical spruce would be loaded with several tons.

Freezing rain often occurs when a lingering arctic air mass gets overrun by warm air ahead of an approaching storm, creating an inversion. The temperature near the ground may be 10 degrees below the freezing point, but several hundred or thousand feet aloft there's a warm layer of above-freezing air. Above that it cools off in the usual manner. Arctic air that settles into the interior valleys of the Pacific Northwest is not easily dislodged by coastal storms, which themselves bring no shortage of melted snowflakes. Some of these locations get glazed an average of 10 or more days a year.

Ice storms are rare in the Southwest, where it just doesn't get cold enough, and in most mountain locations, where suitable temperature inversions don't happen. Rime ice, a close relative of glaze, is more common in the western moun-

RIME—A night of fog at subfreezing temperatures can leave thick deposits of rime on trees, such as these atop Capulin Mountain, New Mexico.

Richard A. Keen

tains. Supercooled fog, rather than rain, freezes directly onto whatever gets in its way, leaving a white coating of ice. Usually rime forms quarter-inch-long spikes that grow especially well on pine needles. In heavy rime storms, though, the thick coating looks like the stuff that used to fill the freezers before the days of frost-free refrigerators.

SLEET

If the low layer of cold air is deep enough, raindrops may freeze solid on their way down. At other times, raindrops are blown back up to the colder levels of a storm, where they freeze and fall back to earth. In either case, these clear pellets of frozen rain are known as "sleet," which bounces when it hits the ground. Although usually less than a sixteenth of an inch in diameter, sleet pellets are round and slippery, and it doesn't take much of the stuff to foul up traffic. Sleet is most common—occurring eight or more times a year—along the Pacific Northwest Coast and in the Bitterroot Mountains of Montana and Idaho.

Sleet pellets, being small and solid, fall faster than snowflakes. They are more likely to reach the ground without melting when the temperature is above freezing. That is why places that rarely see snowflakes do get to see sleet once·in a while.

HAIL

The last member of our precipitation potpourri is one of the nastiest—hail. Hail does millions of dollars of damage to crops every year, and storms have driven unlucky farmers out of business in less than 10 minutes. People have been killed by hail, although, fortunately, these cases are rare. Cattle have less opportunity to find shelter when hail strikes, and their casualties over the years have been much higher. Damaged roofs, broken windshields and dented automobiles all bear testimony to the destructiveness of hail.

Hail is yet another form of refrozen raindrops. It begins just like sleet—with rain blown back up into colder parts of the cloud. With hail, though, the upward winds are much stronger. A frozen drop may spend enough time high in the cloud to gather a frosty layer of ice crystals; when it falls back down it picks up water droplets. The growing pellet may then be blown back up a second time. When the coating of water droplets freezes, the pellet picks up another layer of frost. This up-and-down cycle can happen 10 or more times before the hailstone becomes too heavy for the winds to keep it up.

Next time it hails, cut one of the stones in half with a warm knife. You'll see concentric rings of white and clear ice—alternating layers of accumulated frost and frozen droplets. The number of layers tells you how many loops the stone made through the storm cloud.

Hailstones can range in diameter from an eighth of an inch up to four inches or more. It takes a mighty powerful updraft to send the biggest stones back up for more ice. The heftiest thunderstorms are in the Midwest. A stone weighing nearly two pounds and measuring half a foot across dented Dubuque, Iowa, in 1882. The impact of a large stone can be lethal, with baseball-sized hailstones hitting the ground at 90 miles an hour, the speed of a major league fastball. Notice what baseball catchers wear to protect themselves from such forceful flying objects! Fortunately, giant hailstones are rare—although not unknown—in the West.

ICE FROM THE SKY

It's amazing how much trouble something as ordinary as ice can create when it comes out of the sky. For example. . .

October 1846—Early snows in the Sierras trapped the wagon trains carrying the Donner party to California. Only half of the 81 pioneers survived and were rescued four months later.

Richard A. Keen   Richard A. Keen

HAILSTONES—A hailstone's appearance tells a lot about how it forms. The transparent stones on the left spent much of their time in the warmer, lower parts of the storm cloud, where they collected liquid water which later froze into clear ice. The white centers of the stones are rime ice gathered near the top of the cloud. Some stones have several layers of clear and white ice, indicating two or more up-and-down trips through the storm. The mothball-like stones on the right are composed almost entirely of rime ice, which collected in the colder parts of the storm cloud.

January 11–13, 1888—The "Blizzard of '88" killed 200 people and thousands of cattle across the plains from Montana and Wyoming to Minnesota. A more famous "Blizzard of '88" struck New England two months later.

February 21–23, 1919—Supercooled fog around Cheyenne, Wyoming, froze directly onto trees and wires, causing $150,000 in damage.

April 14–15, 1921—A 32-hour storm left 95 inches of snow at Silver Lake, Colorado, in the mountains northwest of Denver. The one-day total of 76 inches set a national record. A sloppy mixture of rain and snow broke trees and power lines in Denver.

January 1–10, 1949—A series of heavy snowstorms struck the West; some wet snow even fell at San Diego. The worst storm was the blizzard of the century in Wyoming and northeastern Colorado. Three days of gales, snow and near-zero temperatures plagued Cheyenne and stranded thousands of cattle out on ranches. The Army came to the rescue of the starving livestock with Operation Haylift.

January 13–14, 1950—On Friday the 13th, a near-blizzard left 21 inches of wind-driven snow in Seattle.

January 1–3, 1961—Freezing fog in northern Idaho coated wires with rime ice eight inches across.

December 12–20, 1967—Two back-to-back, slow-moving storms sauntered across the Southwest, blanketing the Four Corners states with massive snowfalls. Seven feet fell at Flagstaff, Arizona, collapsing roofs and farm buildings. A repeat of Operation Haylift saved the lives of many people and cattle on the Navajo reservation in northeastern Arizona. San Diego saw a light fall of graupel, and even Yuma, Arizona, had a few flakes.

Richard A. Keen

BLOWING SNOW—A ground blizzard in the Rockies. Winds of 20 m.p.h. can cause drifting of snow but it takes gusts of 30 m.p.h. or more to lift the blowing snow to eye level or higher.

December 30–January 3, 1968–69—Moist Pacific air streaming over a record cold air mass dripped five days of freezing rain across much of Oregon. Falling on top of earlier snows, the ice blocked roads and downed power lines, closing businesses and stopping traffic around Portland for several days.

June 24, 1972—Four inches of snow at Paradise Ranger Station, 5,427 feet up Washington's Mount Rainier, topped off the snowiest winter ever recorded in the United States. Total snowfall since September 1971 measured 1,122 inches, or nearly 94 feet.

Winter 1976–77—Here's an example of what a *lack* of snow can do. An extremely dry winter left many of the West's ski slopes bare, and skiers found other things to do. In Colorado, a 40 percent drop in visiting skiers cost the state several hundred million dollars. Four years later, another "snow drought" had similar results.

July 30, 1979—Softball-sized hail descended on Fort Collins, Colorado, killing a three-month-old baby—the first American hail fatality since 1930.

December 24–25, 1982—The "Christmas blizzard" shut down Denver with two to four feet of wind-driven snow. The city paid $4 million to clear the snow from its streets and airport.

June 13, 1984—Hailstones, some as large as grapefruit, pummeled Denver for 3 hours, punching holes in roofs and reshaping automobiles. Total damage was nearly half a billion dollars.

HOW MUCH SNOW?

Snow lovers and snow haters alike have no trouble finding their favorite climates somewhere in the West. At the snowy end of the scale, Paradise Ranger Station in Mount Rainier National Park, Washington, receives an average of 578 inches per year. Three winters—1955–56, 1970–71 and 1971–72—each dumped more than 1,000 inches of snow at Paradise. A close second, Oregon's Crater Lake averages 541 inches a year, with 879 inches falling in the winter of 1932–33. Other places that have seen more than 800 inches of snow in a winter are Tamarack, California, with 884 inches in 1906–07 and Wolf Creek Pass, Colorado, at 837 inches in 1978–79.

At the other extreme, most of the coastal cities of California have seen just one or two snowstorms in the two- to four-inch range, while Phoenix, Arizona, has on two occasions suffered one-inch snowfalls. Two cities of the West—Yuma and San Diego—have never had a snowfall deep enough to measure. In the past century, flakes have been sighted at Yuma three times, in 1932, 1937 and 1967. However, flurries have struck San Diego only twice, in 1949 and 1967. San Diego is the most snow-free city in the West, although by a very thin hair. Don't forget, however, that Key West, Florida, has never seen a flake of snow.

AVERAGE ANNUAL SNOWFALL IN THE WEST (in inches), 1899–1938

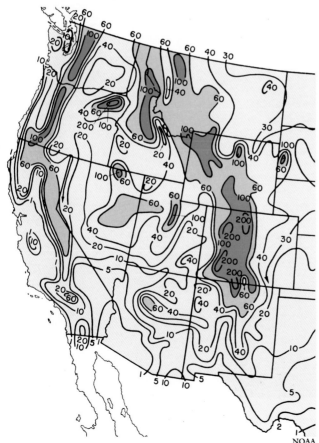

NOAA

41

Alan Moller

A "BOLT FROM THE BLUE"—Positive lightning flashes across clear sky, from the top of the thundercloud to the ground several miles from the storm.

T H U N D E R S T O R M S

Tall, dark columns of thunderclouds seem to boil up through flatter layers and climb into the upper atmosphere. . . . Electric flashes of lightning brilliantly illuminate from within the thunderclouds that generate them. The huge clouds seem to light up instantly and magnificently like enormous bulbs, and a single lightning bolt often seems to trigger a chain reaction of flashes from cloud to cloud so that the lightning appears to be walking its way for hundreds of miles across the darkened earth.

—Joseph P. Allen, *Entering Space* [2]

Whether from the perspective of an astronaut orbiting on board the *Columbia*, or of the rest of us stranded here on earth, few phenomena of nature are more impressive than lightning. Lightning and its inseparable offspring, thunder, are the necessary attendants of thunderstorms, and combined they give the West some of its deadliest and most beautiful weather.

As with almost every other kind of weather, thunderstorms range from extremely rare to commonplace in different parts of the West. In the Rockies of New Mexico and Colorado, thunderstorms strike on 70 or more days a year, while San Francisco averages but two annually. Like snow and heat waves, thunderstorms are seasonal, but even this differs with location—most of the West's thunderstorms rumble during summer; along the coast, winter is the thunderstorm season.

Western thunderstorms are, in general, not as severe as their larger midwestern cousins. That is not to say, though, that severe weather has never visited the West. Tornadoes, hail, gusty winds and downpours have struck every state of the West, and at times with terrible losses of human life and property.

CLOUDS AND CONVECTION

Clouds need water if for no other reason then just to be seen. Clouds are liquid droplets of water condensed from the vapor contained in rising currents of air. However, thunderstorms need water for more than droplets—it's the latent heat they're really after. When water vapor condenses in rising air currents, its latent heat is released, warming the air surrounding the new cloud droplet. Like a bubble in a pool, this warmer—and lighter—air rises more swiftly, leading to even more rapid condensation. Meanwhile, increasing amounts of fresh air drawn into the base of the cloud keep the condensation going. This process of rising, condensation heating and faster rising currents is called "convection." Without it, there would be no thunderstorms.

Even the meanest thunderstorm needs something to get it going. Nothing happens until air starts rising. For most western thunderstorms, updrafts begin with air heated by the sunlit ground below. Some storms, though, start with the air currents in different sections of cyclones, especially where the cold front wedges under warmer, moist air and shoves it upward.

There's plenty of water in the air and no shortage of sunlight and fronts to get those currents going. Nonetheless, thunderstorms are relatively infrequent. Even those places that get thundered on 70 days a year *don't* get thundered on the other 295 days. The average storm lasts an hour or so, which means only about 100 hours—or four days—out of the 70 are actually thundering. So even in the thunderstorm capitals of the West, it's actually thundering only 1 percent of the time during an average year. Most people spend more of their lives in the shower than they do hearing thunder.

Clearly, atmospheric conditions have to be just right for thunderstorms to develop. The necessary condition is called "instability," the opposite of "stability." To physicists, a stable object (or situation) is one that returns to its original position after being moved (or changed). A bowling ball sitting in a ditch is stable—give it a shove, and it will roll back to where it was. A bowling ball balanced on a roof is unstable—give it a kick and

off it goes.

Like a bowling ball, the atmosphere can be stable or unstable. Cold air in a valley—an inversion—is stable, since the air becomes warmer with height. Lift the cold air off the ground, say, 100 feet, and it is surrounded by warmer, lighter air. Let go, and the cold air returns to the valley floor. With stable air, those little rising currents don't get very far.

With unstable air, the temperature drops off fairly rapidly with height—about 25 degrees per mile. If ground-level air is lifted, it expands and cools as the pressure gets lower. Despite this cooling, however, the lifted batch of unstable air remains warmer than the air around it, and keeps going up. Instability is even stronger if there's water vapor in the air to release its latent heat. If you give an upward kick to part of an unstable air layer, that part continues to soar upward in a relatively warm stream.

You can't have rising currents of air everywhere, or the lower atmosphere would soon run out of air. That's why thunderstorms are so scattered. The clear spaces between thunderstorms, which are much wider than the storms themselves, are where the air is sinking back to the ground. With air bubbling up here and sinking over there, afternoon thunderstorms are part of an extremely complex pattern of up-and-down currents.

THUNDER IN THE HILLS

Now we know how to cook up a thunderstorm. All we need is some unstable, moist air and something to kick it with. Clearly, this doesn't happen very often along the West Coast, especially along that stretch between the coastal ranges and the sea. In Washington and Oregon, Quillayute and Astoria average only eight thunderstorms a year. In California, there is an average of only five thunderstorms at Eureka and two at San Francisco. Inland a little, Stockton and Sacramento average but one thunderstorm a year. Farther

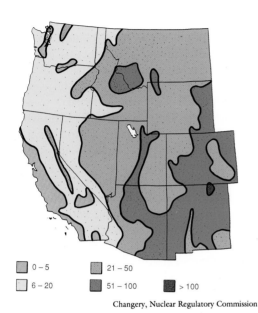

0 – 5		21 – 50	
6 – 20		51 – 100	
		> 100	

Changery, Nuclear Regulatory Commission

THUNDERSTORM FREQUENCY—Number of days with thunderstorms across the West.

The farther inland you go, the more thunderstorms there are. In the Sierras there may be 20 or 30 thunderstorms in a year; the plateaus and mountains of Arizona and Utah, as well as the High Plains from eastern Montana to New Mexico, may see 50 or more. But nobody hears more thunder than the residents of the Rockies, from New Mexico into Colorado. The thunderstorm capital of the United States lies in the Sangre de Cristo range west of Raton, New Mexico, where storms rumble an average of 110 days a year. The Tampa Bay area of Florida, considered by some to have the most thunderstorms, averages about 100.

Notice that the West's most thunder-prone locations are in the mountains and that the highest mountains—the central Rockies—have the most storms of all. Even within mountainous areas, there are more storms in the highlands. At 9,000 feet elevation, I've recorded an average of 73 thunderstorm-days per year, while Denver, 28 miles away and 4,000 feet lower, averages only 41. That's a nearly two-fold difference between places within sight of each other!

Why do mountains get so many thunderstorms? If you live near a mountain, you've probably noticed that the first puffy clouds of the day form over the mountains and that the mountain clouds grow faster than their lowland counterparts. Mountains get off to an early start because their air is less stable than in the valleys. Overnight, cold air drains from the highlands to the lowlands, leaving the valleys full of cold, stable air. By the time the sun heats up this stable layer, clouds are already billowing over the hills.

LIFE OF A CUMULONIMBUS

Like humans, thunderstorm clouds change their size and shape as they grow, reach their prime, and then get ready to check out. The first clouds to form are white, cottony puffs called *cumulus* clouds, a Latin word meaning 'pile' or 'heap.' Who among us has never seen fleeting

south, things pick up again; thunderstorms strike Los Angeles and San Diego about three times a year. Unlike thunderstorms in every other western region, coastal storms prefer the winter months.

The ocean-chilled air along the California coast is the most persistently stable layer in the country. Two thousand feet up, temperatures may be 30 degrees warmer than at sea level. This strong inversion lasts from spring until fall and is very effective at keeping thunderstorm activity down. The coastal inversion is so strong that its cool air pours through the Golden Gate to suppress thunderstorms in the Central Valley. The inversion isn't quite so strong farther north, but is still strong enough to do its job. In winter, the air aloft isn't nearly as hot, and the coastal inversion weakens. At times (two to eight times a year, to be exact), a winter cold front is strong enough to get things convecting, and San Franciscans marvel at the flashes in the sky.

45

forms of sheep, giraffes and dragons in these heaps of water droplets? As they turn more vapor into droplets, cumulus clouds grow into billowing cauliflowers named "towering cumulus."

Three or four miles up (lower in the winter) lies the freezing level. Above this line, the air is colder than 32 degrees, and clouds that go higher may freeze. It usually takes a while for an ascending droplet to find a freezing nucleus, so for several thousand feet above the freezing level, the cloud is made of subfreezing, but still liquid, water droplets—supercooled water. The cloud may reach 30,000 feet before its droplets start to freeze.

With a frozen top, the cumulus cloud becomes a different animal. The cloud top loses its solid, cauliflower appearance, and takes on the stringy, fuzzy appearance characteristic of ice clouds. High winds aloft may blow the ice crystals away from the main cloud; the flat, spreading tops of ice are often called "anvil" clouds.

The changing appearance is a sign that something more meaningful is happening in the cloud. In the moist interior of the cloud, ice crystals grow rapidly into snowflakes, which fall and melt into rain. *Nimbus* is the Latin word for rain, and our cloud has matured into a *cumulonimbus*, meaning, loosely, a raining heap!

While cumulus and towering cumulus clouds are growing, air currents all head upward. But as ice crystals form, some of this air starts coming back down, bringing with it snow and rain. As the downdrafts reach the ground, they spread out, bringing us the first gusts of cool air that so often are followed by sudden rains. On the ground, a storm has begun.

If a raindrop loses its downdraft, it may

LIFE CYCLE OF A THUNDERSTORM—During its lifetime, which may last one or several hours, a thunderstorm goes through three stages: the towering cumulus stage, with updrafts building the cloud; the mature stage, with updrafts feeding moisture into the cloud, to fall as rain in downdrafts; and a dissipating stage, in which the supply of fresh moisture is cut off and the water in the storm is "raining out."

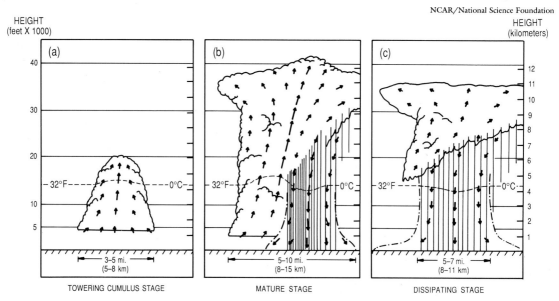

NCAR/National Science Foundation

Ronald Holle

CUMULUS CLOUDS—"The clouds must look like many sheep before the rains will come."
—Navajo proverb

Richard A. Keen

TOWERING CUMULUS—The characteristic "cauliflower" appearance indicates a rapidly growing cumulus cloud.

David O. Blanchard

MATURE CUMULONIMBUS CLOUD—A spreading "anvil" cloud of ice crystals tops this thunderstorm.

David O. Blanchard

MAMMATUS CLOUDS—These pouch-like formations hanging from the side of a cumulonimbus are called "mammatus" clouds, from their remote resemblance to the breasts of female mammals. The cloud formations are caused by downdrafts of cool, moist air.

Richard A. Keen

SMALL THUNDERSHOWER SEEN FROM THE SUMMIT OF WYOMING'S GRAND TETON PEAK—The white streamers falling from the base of the cloud are probably made of graupel (or snow pellets) which melt into rain at lower levels. As is typical of many western thundershowers, much of the precipitation evaporates before reaching the ground.

Richard A. Keen

DISSIPATING STORM—Smaller yet is this fading storm. All of the precipitation is evaporating far above the ground.

Richard A. Keen

CIRRUS CLOUDS—The cycle of up- and downdrafts also happens high in the atmosphere. These cirrus clouds are actually high-altitude showers, with snow crystals forming in the cottonlike puffs and falling in thin streamers. The falling streamers of snow trail behind the puffy clouds, which are being blown to the left by high winds at these altitudes (30,000 to 40,000 feet).

Richard A. Keen

WISPY CIRRUS CLOUDS—Even these clouds started out as eight-mile-high snow showers. Note that many of the wisps have small tufts at their upper right ends. Some of the tufts have already dissipated, leaving only a trail of falling ice crystals.

blow back up above the freezing level to become hail. The up- and downdrafts can approach 100 miles an hour in a mature cumulonimbus cloud. Loaded with hail, these vertical winds give pilots good reason to avoid thunderstorms. Even on the ground, hail may arrive on hurricane-force winds, giving the earthbound equally good reason to get out of the storm.

At some point, the supply of fresh, warm, moist air runs out, and the thunderstorm begins to fade. Inside the cloud, the updrafts weaken, shrink and finally cease, leaving sinking currents of rain- and snow-filled air. The cloud's edges begin to evaporate, and it takes on a ragged appearance. Above, the ice-crystal anvil cloud may separate and blow off with the high-level winds, while below, light rain falls out of a disappearing cloud. The entire cycle of growth, maturity and decay may take two or three hours, and by the fourth hour there may be no visible remnants of a once mighty storm!

Most western storms are started by solar heating. While the first cumulus generally appear over the mountains by late morning, clouds may pop up within an hour or two of sunrise if the air is especially moist and unstable. On dry, stable days, they may never appear. Several studies of the timing of daily thunderstorms have found that, in general, peak thunderstorm activity—when the most storms are going on at the most places—is around 4 p.m. In the high mountains, peak storm activity is earlier, around 1 p.m., while the lowlands see most of their storms lingering until midnight or later. Veteran mountain climbers know these patterns well and know the advantages of starting their climbs early.

These statistics reveal an important pattern. Thunderstorms develop first over the mountains and, as afternoon wears on, move out to the foothills. The mountains may already be clearing while the foothill storms are just beginning. Some storms move onto the plains and into the valleys during evening, but most die off by 10 p.m. If the lowland air is especially moist and unstable, however, storms may continue until dawn.

Rarely, though, can individual storms be traced from the mountains to the lowlands throughout the course of the daily cycle. What happens is more like a moving "wave of activity," within which storms live their several-hour lives and are replaced by new storms. These waves can travel considerable distance. The wave of thunderstorms that begin over the Colorado Rockies around 2 p.m. can sometimes be traced to eastern Kansas, 400 miles east and 12 hours later. There have even been occasions when the thunderstorms continue well into the next day, by which time they're approaching the Mississippi River. This pattern seems to hold for all parts of the West, from the Rockies to the Sierras, although it's most pronounced where the storms are most frequent.

Why does this wave of thunderstorms happen at all? It is, essentially, the thunderstorms' way of looking for greener pastures. Being generally cooler and drier, mountain air usually doesn't contain a whole lot of moisture to feed thunderstorms, and those first mountain storms soon run out of fuel. However, their cold downdrafts spread across the foothills and, acting like small cold fronts, start another generation of storms. These new storms use up their local moisture supply, start off a third generation and so on. Over and over this cycle continues, as new storms feed on untapped moisture supplies ever farther from the mountains. If the mountain storms clear off early enough, the sun may kick off another round of thunderstorms; sometimes there are several waves during the course of a day. What hour of the day has the most thunderstorms at your home? Keep records for a year and you'll get a pretty good idea. It's an easy, interesting and cheap weather project.

The vast majority of western thunderstorms are started by sunlight. Sometimes, however, an advancing front—particularly a cold front—gives the air its upward shove. Thunderstorms pop up here and there along the front, sometimes merging into a line extending tens or even hundreds of miles along the length of the front. The approach of a "squall line" can be an imposing spectacle, with a wall of dark, thundering clouds stretching from horizon to horizon. Squall lines are rare everywhere in the West, but along the coast—where solar-powered storms are virtually nonexistent—most of the few thunderstorms that do occur are set off by fronts.

KING LEAR'S QUESTION

First let me talk with this philosopher—what is the cause of thunder?

—Shakespeare, *King Lear*

The Norse had an answer—the mischievous Loki forged bolts and gave them to Thor, who pitched them earthward. When Thor rode his chariot, thunder rumbled across the heavens. However, Vikings saw even fewer thunderstorms than Californians, so perhaps we shouldn't consider them the experts. Their explanation was as good as any, though, until Benjamin Franklin performed his legendary lightning experiment in 1752.

We all recall how Franklin flew his kite into an approaching thunderstorm. Contrary to popular belief, his kite was *not* struck by lightning; if it had, Franklin's signature would have never appeared on the Declaration of Independence. The developing storm was generating a lot of static electricity, though, and sent fibers on the kite cord standing on end. Suspecting the presence of electricity, Franklin touched a metal key tied to the end of the cord. His suspicions were brilliantly confirmed.

Two centuries after Franklin's experiment, the source of all this electricity is still a mystery. A recent pamphlet on lightning by the National Oceanic and Atmospheric Administration (NOAA), our nation's largest weather research outfit, noted that "no completely acceptable theory explaining the complex processes of thunderstorm electrification has yet been advanced. But it is believed that electrical charge is important to formation of raindrops and ice crystals, and that electrification closely follows precipitation."

In other words, electricity appears to be produced by the freezing of supercooled water droplets. For some reason, when some drops freeze, the ice crystals take on a positive charge, and the remaining droplets become negative. Not only does the ice crystal anvil formation atop a cumulonimbus portend rain, it also means lightning and thunder. That's why cumulonimbus clouds are more popularly known as thunderclouds.

At this point of freezing, the electrical charges in the cloud separate, with the positive charges gathering near the icy cloud top and the negative charges gathering in the lower parts. Under normal conditions, the ground is also negatively charged. Now, however, the concentration of electrons in the lower cloud repels the negative ground charge (like charges repel; opposites attract), leaving a positively charged ground for several miles around the cloud base.

As the charges gather, voltages build up to as high as 100 million volts within the cloud and between cloud and ground. Air is a terrible conductor, meaning that it is fairly effective at holding electrical charges apart. In clouds, air can separate voltages at the rate of 3,000 volts per foot, or 15 million volts per mile. However, enough is enough, and when a thundercloud's 100 million volts show up, something has to give.

Sparks fly when cloud voltage reaches the breaking point, but their flight is not a direct one. One hundred feet at a time, a "leader stroke" makes its way downward from the base of the cloud. Too faint to be seen, the leader ionizes the air along its path, meaning that this air can now conduct electricity. It pauses for a few millionths of a second, then jumps another 100 feet. After a hundredth of a second, the leader stroke arrives within a few hundred feet of the ground. At this point, similar leader strokes stream upward from hilltops, trees, antennas, golfers and the like to join the downward leader. This meeting completes an electrical circuit between the cloud and the ground.

With cloud and ground connected with a conducting "wire" of air, the electrical charges go into action. A "return stroke" shoots up the leader path at one-sixth the speed of light. Sometimes, return strokes follow each of the branching leader channels up from the ground, to join in a single massive stroke several hundred feet up. The intense electrical currents, concentrated in a path a few inches across, heat the air almost instantaneously to tens of thousands of degrees—several times as hot as the surface of the sun! We see this glowing channel of hot air as the familiar and spectacular flash of lightning.

The return stroke doesn't last long; in less than .00005 seconds, it's over. If there's still some electrical charge left in the cloud, another leader stroke may head down the path of ionized air left over from the first stroke. When it reaches the ground, another powerful return stroke heads back up the channel. Every .05 seconds, this cycle repeats itself until the cloud is discharged. There's often enough electricity to make two or three separate strokes. Sometimes there are 10 or more, giving some lightning a flickering appearance. Properly speaking, these repeated strokes are called "flashes" of lightning. The popular term "bolt" is not precisely defined and is considered colloquial by meteorologists.

This sudden heating literally explodes the air all along the length of the flash, and we hear the detonation as thunder. Thunder can usually be heard eight or 10 miles away from the lightning that produced it, but there are some reports of the sound being heard 20, 30 and even 70 miles away. You can measure the distance of lightning by counting the seconds until you hear the thunder. The boom of thunder travels at the speed of sound, .2 miles per second, so by dividing the elapsed time by five you'll get the distance in miles. Next time your ears are rattled by a thunderbolt, just remember that less than 1 percent of lightning's energy goes into making noise.

Lightning that zaps from cloud to ground is called, with the usual scientific honesty, a "cloud-to-ground" flash. Lightning also jumps between differently charged regions within a cloud, such as from the lower regions to the icy top. This "inside-cloud" lightning is actually more frequent than the cloud-to-ground variety. Lightning connecting differently charged parts of *different* clouds is called, as you might expect, "cloud-to-cloud."

These three types of lightning are well known to meteorologists, and each even has its own abbreviated code name for transcribing onto weather reports. Recently, yet another kind of lightning has been documented, and it may be the nastiest kind of all. Recall that cloud-to-ground

FOUR KINDS OF LIGHTNING—All forms of lightning connect oppositely charged regions inside clouds or on the ground.

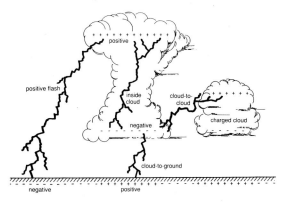

Richard A. Keen

INTENSE ELECTRICAL STORM—Two kinds of lightning are seen here—inside-cloud and cloud-to-ground—as an intense electrical storm rakes the eastern plains of Colorado.

flashes go from a negatively charged cloud base to the positively charged ground directly underneath. Meanwhile, several miles away from the storm, the ground is *negatively* charged. This sets the stage for cloud-to-ground lightning to jump from the positively charged cloud top to the negative ground. Since these flashes discharge from the positive region of a cloud, they are often called "positive" flashes.

Positive lightning flashes are not just negative flashes going the opposite direction. For one thing, the distances they traverse can be enormous. From a 40,000-foot cloud top to the ground five or more miles from the storm, a single flash may extend 10 or even 20 miles—five times the length of a typical negative flash. Positive flashes are often among the last thrown off by an about-to-die storm. They seem to prefer winter thunderstorms—perhaps because the positively charged icy parts of the cloud are closer to the ground.

There is a treacherous quality to these positive flashes. Coming from a storm miles away, they may appear to strike out of a clear sky. I once saw one of these flashes come from a storm passing several miles to the south. The lightning left the storm, passed across blue sky directly overhead, and connected to a ridge off to the north. There are doubtlessly some unfortunate souls who took the proper precautions during a thunderstorm, only to be struck by a "bolt from the blue" as the storm moved on and the sun was shining.

There's an old adage that "lightning never strikes twice in the same place." I can assure you that this is not so, having once been in a truck that was struck three times in five minutes, and on another occasion having watched a utility pole endure three strikes in 20 seconds. In neither case was there any damage! These six strikes illustrate how much lightning there actually is. A lightning

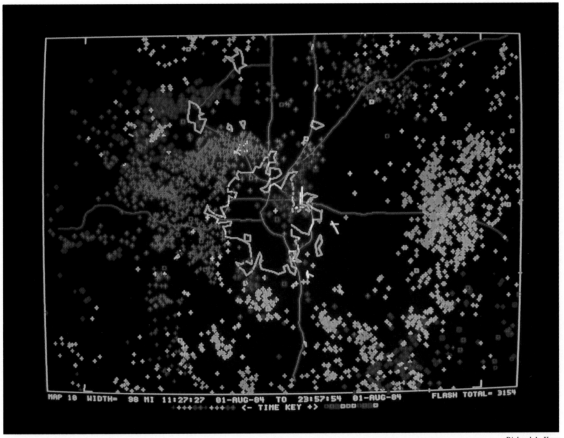

ONE DAY'S LIGHTNING STRIKES—On August 1, 1984, an electronic lightning detection system located 3,154 cloud-to-ground lightning strikes centered in the Denver area. On this computer-generated map (photographed from a video screen), irregular light blue lines mark the areas of cities and towns, while the purple lines are major highways. The varicolored crosses mark the locations of the more common negative cloud-to-ground flashes, while the small squares mark the rarer positive strikes. The detection system consists of directional antennas which pick up ground waves from the lightning.

detection system over the western states has been picking up more than a million cloud-to-ground strikes per summer, or about one every square mile. Some mountain areas have been bombarded with five or more per square mile during the few short months of summer. With that much lightning to go around, some is bound to find a target

that's been struck before. If there's any truth to the adage, it's because some targets just aren't there after the first strike.

It's easy to describe the power of lightning with some very large numbers. The *instantaneous* peak rate of energy usage may exceed a trillion watts, equivalent to the *average* consumption rate

59

of the entire United States. But lightning strokes are incredibly brief, ranging from millionths to thousandths of a second long, and the *total* electrical energy expended by an average lightning flash is several hundred kilowatt hours. It depends on where you get your electricity, but from where I get mine, an average flash would cost about $50.

It is a bit more meaningful to measure the power of lightning in terms of what it does to its targets. Perhaps the most impressive and lasting monuments to lightning are glassy masses of melted and solidified sand called "fulgerites." Fulgerites, which may be several inches across and 20 to 40 feet long, often mark the path of electricity as it skims along the ground. Their total weight can run up to several hundred pounds. Sand melts at 3,100 degrees; now consider the energy it takes to heat 200 pounds of sand to that temperature, all in a fraction of a second!

It is no wonder that lightning is one of the West's deadliest weather phenomena. Over the past 25 years, lightning has killed an average of 100 Americans annually, nearly as many as tornadoes and hurricanes combined, and injured 250 more. Colorado and New Mexico led the 11 western states in lightning casualties, while Washington has had but one fatality in the past 30 years.

Many of those killed and injured by lightning were engaged in outdoor activities, such as boating, golfing and baseball. Hikers above timberline are particularly vulnerable; atop a mountain, a climber may be the most tempting lightning target in the state. However, even some indoor activities can be dangerous during a thunderstorm—people talking on the telephone have died in mid-conversation when a nearby strike came "over the wires." Lightning is beautiful and spectacular, but it also deserves a great deal of respect.

LIGHTNING'S GREATEST HITS

July 22, 1918—A single lightning strike killed a flock of 504 sheep grazing in Utah's Wasatch Mountains.

April 7, 1926—Lightning struck an oil depot in San Luis Obispo, California, setting it aflame. When the fire was extinguished five days later, it had killed two people, burned 900 acres and consumed 6 million barrels of oil.

July 25, 1945—An early morning thunderstorm forced an hour and a half delay in the first test of an atomic bomb near Alamogordo, New Mexico. For obvious reasons, technicians were reluctant to perform adjustments on the device while lightning was striking nearby.

August 15, 1967—Lightning touched off the 56,000-acre Sundance fire in Idaho's Selkirk Mountains. Intense flames generated tornado-like whirlwinds that blew burning trees around like matchsticks.

August 23–24, 1970—Dry thunderstorms ignited 100 separate fires in Washington's Cascade Mountains. The fires consumed 100,000 acres of forest and rangeland before they were finally controlled a week later.

July 12, 1984—A "bolt out of the blue" from a storm three miles away struck and killed a golfer in Tucson, Arizona.

July 6, 1985—Most of Utah was blacked out for several hours by a lightning strike to a generating station in Salt Lake City. The exploding transformer sent flames 200 feet into the air.

July 27, 1985—Lightning struck five hikers atop Half Dome in Yosemite National Park, California, killing two of them. The same day, lightning killed another hiker in Sequoia National Park, 100 miles away.

VIOLENCE IN THE SKIES

Fierce fiery warriors fight upon the clouds,
in ranks and squadrons and right form of war.

—Calpurnia to Caesar,
in Shakespeare's *Julius Caesar*

Caesar's wife saw thunderstorms as omens of his impending death. We see thunderstorms as welcome bringers of rain and relief from summer heat. That relief, however, is not always benign, and the ominous dark clouds of an approaching thunderstorm may still presage destruction and even death.

HAIL AND HIGH WATER

Thunder and lightning aren't the only mischief thunderstorms can make. Thunderstorms can also be prodigious rainmakers, and when too much rain falls too fast in too small a place, the result is a flash flood. Lightning is quick, tornadoes are awesome, but no weather-related disaster kills more westerners than flash floods.

The first requirement for a thunderstorm to cause a flood is the obvious one—that it produce an exceptional amount of rain. However, this is not enough. Most thunderstorms move along at a sprightly 10 to 40 miles an hour, which means the storms last half an hour or so at any one place. Flash flood storms, on the other hand, stand still for hours. Tremendous amounts of rain can dump on one valley, while the next valley—within earshot and hearing thunder throughout—gets a sprinkle.

Why do a select few thunderstorms stand still while most of their peers move? The motion of thunderstorms is not a simple process. To some extent, it is controlled by upper-level winds, which continually nudge along the top portions of thunderclouds. Thunderstorms may also move because they're growing faster on one side than on the other, and the cloud mass shifts in that direction. Remember the wave of thunderstorms that sweeps off mountains? This is much the same thing—cold downdrafts spreading out at the base of the storm

61

SKYWATCH

CHEYENNE—For several hours during the evening of August 1, 1985, an intense thunderstorm sat motionless over Cheyenne, Wyoming. Up to seven inches of rain drenched the city, causing street and stream flooding that took 12 lives and caused $65 million damage. Two-inch-diameter hailstones accompanying the rain were washed into piles eight feet deep along roads and five feet deep in basements.

Timothy Kittel, Colorado State University

Just after sunset, the Cheyenne storm presents an impressive spectacle photographed from Fort Collins, Colorado, 40 miles south. As often happens in flash flood thunderstorms, frequent lightning flickers inside the cloud during this four-minute time exposure.

Michael Mee, Federal Emergency Management Agency

Floodwaters washed hail and cars into roadside heaps.

give moist air an upward shove, and new towers of cumulus build up on the side of the thundercloud. The building cumulus towers tap fresh supplies of water vapor, keeping the thunderstorm alive.

It may seem contradictory to have a stationary thunderstorm produce lots of rain, since cold air collecting at the base of the storm should cut off its moisture supply. But what if a wave of thunderstorms tries to move, say, east into the face of a westward blowing wind? If the speeds match, the wave stands still. The thunderstorms also stand still, while easterly winds feed fresh, moist air into their hungry jaws.

Thunderstorms are more likely to stay put if there's a mountain nearby, since sloping terrain encourages the storms' growth by forcing moist wind upwards. That's why the West's heaviest thundershowers are in the mountains. The problem arises because much of the population of mountainous areas lives in the valleys. Of course, you don't need a mountain to make a flash flood, since places like Kansas City and Houston have been struck. But in the West, a thunderstorm sitting over the head of a valley with a ready supply of moist air is the prime ingredient for a flash flood.

Not all of the water falls as rain. We saw in the last chapter how strong updrafts can blow raindrops back into the frozen parts of the cloud. These frozen drops grow into hail. There's not much more to say about hail here, except that the stronger the updraft, the bigger the hailstones can grow. Strong updrafts also feed water vapor into the cloud at a faster rate, stepping up the production of raindrops. So hail and high water very often go together.

GUST FRONTS

At its peak of vigor, a healthy thunderstorm has both updrafts and downdrafts. Each has its role: upward currents lift moist air to the condensation level, and downward currents carry the load of rain and hail earthward. The core of the

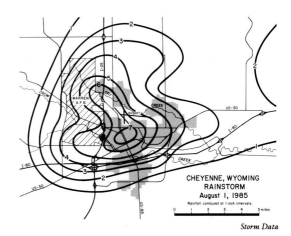

CHEYENNE, WYOMING
RAINSTORM
August 1, 1985
Rainfall contoured at 1 inch intervals.

0 1 2 3 4 5 miles

Storm Data

DISTRIBUTION OF RAINFALL FROM THE CHEYENNE STORM—Most of the heavy rain was localized within a five-mile radius of downtown Cheyenne, where up to seven inches fell.

downdraft may be several miles wide, but as it approaches the ground it has no choice but to spread out horizontally. The leading edge of the expanding outflow of cold air is known as a "gust front," which may spread over an area 10 or 20 miles wide.

Gust fronts are often more than breezes; they've flattened trees and buildings, sunk boats and sent cars off the road with their 100- mile-per-hour winds. They have broken up man's works, from picnics to airplanes. In the next chapter we'll see how gust fronts can spawn small whirlwinds strong enough to be called tornadoes.

The passage of a gust front is marked by a sudden rush of wind and a drop in the temperature, often by 20 or 30 degrees in as many minutes. If the ground is dry and dusty, the gust front may actually become visible as a 1,000-foot-high, dust-laden wall of air. Dusty gust fronts are known as "haboobs," a word which, all jokes aside, comes from the Sudan, where such things are commonplace. They are also common in the desert regions of the West, and some spectacular ones have passed across Phoenix. In moister climates such as the High Plains, ragged bands of low clouds may form along the gust front. Sometimes these low cloud bands form in layers and sometimes they rotate like a log rolling on the ground; the respective names are "shelf" and "roll" clouds.

To meteorologists, any difference of wind speed or direction between two places is called a "wind shear." Usually it's nothing to worry about, but in recent years the sharp wind shears along gust fronts have crashed a mounting number of airplanes, in some cases killing 100 or more people. These tragedies have led to several research projects, such as the JAWS (for Joint Airport Weather Study) project near Denver in the summer of 1982. JAWS was followed a few years later by CLAWS, meaning, of course, Classify and Locate Airport Wind Shear.

DOWNBURSTS

Using such sophisticated equipment as Doppler radar, which can measure wind speeds throughout the interior of a storm, along with more basic techniques such as observers in the field with binoculars and cameras, researchers have discovered a diminutive downdraft that may cause the most dangerous wind shears. The narrow streams of air are less than a mile across and may only span 100 yards or so. As the stream hits the ground, its 50-mile-per-hour downdrafts may turn into spreading winds of 100 m.p.h. or more, sort of like water from a faucet splattering onto the sink. The sudden, nearly explosive winds that result have earned these intense downdrafts the name "downbursts."

Strong downbursts have felled trees in radial patterns, with all the lumber pointing away from the center; sometimes the damage area is only a few hundred feet across! Park chairs have been sent flying, while outside the downburst, picnickers watch in disbelief. Sometimes downbursts are visible as expanding rings of dust

picked up by the winds. Look fast, though—the typical downburst lasts but two or three minutes.

The brief but high winds of a downburst can down trees and houses, but their greatest havoc is wreaked upon airliners. Airplanes are kept in flight by the lift of the air flowing over their wings; the faster the airspeed over the wings, the greater the lift. Every plane has a specific airspeed, called the "stall speed," below which there's not enough lift to keep the plane flying. Airspeed is the speed across the wings, and is the combination of the plane's ground speed plus the speed of a headwind (or minus the speed of a tailwind).

Airplanes are most susceptible to downbursts right after takeoff or just before landing. At these times airplanes are closest to the ground, where the strongest downburst winds occur. It is also when planes are flying the slowest, and may be barely above the stall speed. Encountering a downburst, a jetliner may see a 30 m.p.h. headwind change to a 30 m.p.h. tailwind in 10 seconds. If the sudden loss of lift occurs at too low an altitude, the plane has no chance to recover. In 1982, a jetliner with 144 aboard was downed by a downburst 30 seconds after taking off from New Orleans; there were no survivors. There have been several incidents of jetliner crashes and near-crashes in the West, but, fortunately, without any deaths.

Downdrafts in thunderstorms are usually accompanied by brief heavy rain. It might then seem odd that most of the downbursts studied in the JAWS experiment did *not* have any rain, and that some of the strongest bursts were on hot dry days. These dry downbursts are excellent raisers of dust, but how do they get their winds? It turns out that dry downbursts begin as rain showers, but the rains never reach the ground. As the rain-laden downdraft drops below the cloud, raindrops start evaporating. Cooled by evaporation, the down–draft air becomes heavier and falls faster—the exact opposite of condensation and convection.

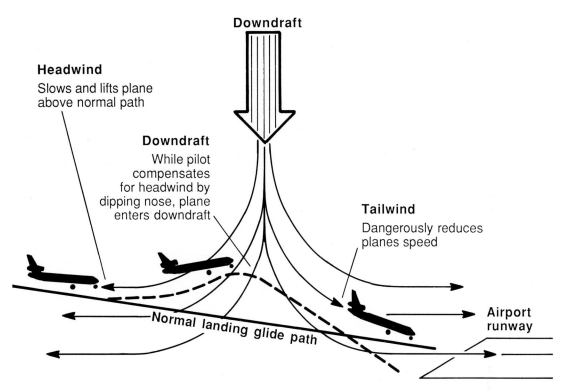

Downdraft

Headwind
Slows and lifts plane
above normal path

Downdraft
While pilot
compensates
for headwind by
dipping nose, plane
enters downdraft

Tailwind
Dangerously reduces
planes speed

Normal landing glide path

Airport
runway

NCAR/NSF

DOWNBURST DANGER TO AIRCRAFT—Air currents spread as a strong downdraft reaches the ground causing rapidly changing winds that can prove fatal to unwary aviators.

Most people have probably seen downbursts in the making, in the form of thin, dark streamers of rain dropping from a cloud. If the shaft of rain disappears before reaching the ground, it's called "virga," and beneath it there's probably a stiff downdraft. If the cloud doesn't make a lot of rain, the downburst may fall as a bubble of cool air; when it lands, its winds might last only seconds.

Downbursts are small, brief and scattered. Some fall from thunderstorms, but others come out of the most inoffensive-looking cumulus clouds. This makes them difficult to detect while

they're happening and virtually impossible to predict even minutes in advance. Fortunately, there are downburst detection devices currently under development that should make landing and taking off a bit safer in the not-too-distant future.

There is one practical step you frequent flyers can take to reduce your odds of being ruined by a downburst. The JAWS experiment found that downbursts in the Denver area are most likely between 1 and 6 p.m., when the sky contains the highest concentration of cumulus and cumulonimbus clouds. Other airports in the arid regions of the West also probably get their downbursts in

David O. Blanchard

TYPICAL WESTERN TWISTER—A small tornado whips the High Plains. The air is too dry for the characteristic funnel cloud to form, and this tornado is visible because of the dust it has picked up. These dust whirls are typical of the small tornadoes that strike much of the West.

Andrew I. Watson

HIGH PLAINS FUNNEL—A classic tornado funnel scours the countryside near Akron, Colorado. The air was moist enough for a complete funnel cloud to form, making this storm more like a midwestern tornado than a typical western twister.

the afternoon. The moral is: fly early!

TWISTERS

There is no doubt that tornadoes are a very special form of weather. Nowhere else does the atmosphere display its power in a more visual and dramatic style; even the most casual cloud watcher will take note of a tornado. But's let's make it clear right here: few westerners ever see a tornado, and fewer still are injured by one. When they strike, it's news.

Like downbursts, tornadoes have powerful winds. In tornadoes, however, most of the air is going up! Tornadoes are picky, too, and happen only when there's a very special set of weather conditions. First, the air has to be moist enough to feed a growing thunderstorm. Next, the atmos-

phere has to be highly unstable, in order to get some extremely strong updrafts going. Third, the lowest layers of air must be slowly turning, like a weak cyclone. Finally, it helps to have a strong jet stream aloft—most tornado-producing thunderstorms are movers.

When a thunderstorm develops in these unusual situations, its powerful updrafts draw in the slowly rotating air. This concentrates the spinning motion, like water swirling faster as it approaches a drain. As the updraft strengthens, the spinning speeds up, until the updraft becomes a narrow, rotating column—a tornado.

In exceptional cases, the whole thunderstorm may start rotating, becoming a parent cloud that may produce tornadoes every few minutes for an hour or more. These awesome storms ravage

SEATTLE TORNADO—The strongest tornado ever to visit the Seattle area passes near the Boeing Aircraft Company plant at Kent, Washington, on December 12, 1969. Tornadoes are extremely rare in the Pacific Northwest but the moist coastal air often allows complete funnel clouds to form.

James Walker, The Boeing Company Archives

the Midwest every year, and have even appeared in the High Plains region. They are rare, however, west of the Rockies.

Tornadoes come in many sizes and shapes. Some midwestern monsters have mile-wide funnels with 300-mile-per-hour winds that may scour the land along a 50-mile path before they dissipate. Tornadoes like this have never struck the West. The "typical" tornado, based on nationwide statistics, has winds of 150 m.p.h. swirling around a 700-foot-wide funnel, and moves at 40 m.p.h. over a five-mile-long path during its five- or 10-minute lifetime. Even this typical tornado is rare in the West.

Most western tornadoes don't even look like tornadoes. The famous funnel cloud comes from the condensation of water vapor inside the rotating column, where the pressure and temperature are lower. In the dry climate of the West, the funnel cloud very often fails to fill the whirlwind. The typical western tornado has a stubby funnel protruding from the cloud and a whirlwind full of dust and debris on the ground. As in any tornado, the rotating tube of air extends all the way from the cloud to the ground, but it's invisible. Some people prefer to call these tornadoes by their diminutive name, "twister."

So the West doesn't get the really big tornadoes, and the small ones often don't even look like tornadoes. Even these small ones aren't that common. Since 1953, the 11 western states have averaged 53 tornadoes per year, only 7 percent of the national total. Of these, 38 touch down on the High Plains from Montana to New Mexico. In an average year, only 15 tornadoes occur west of the Continental Divide.

By American standards, tornadoes are rare in the West. However, three-fourths of the world's twisters touch down in the United States, making the tornado as American as the hot dog and apple pie. By global standards, the yearly quota of 15 twisters west of the divide turns out to be fairly impressive. For comparison, the entire continent of Australia has reported an annual average of 14 tornadoes, and relatively tornado-prone Japan has averaged 11 twisters per year. Probably the only countries that see more tornadoes than the West are Great Britain and Nevada-sized New Zealand, each averaging 25 or so a year.

Small, rare and invisible they may be, but no tornado should be sneezed at. In the seven decades since 1916, tornadoes have killed 52 westerners. Half of them were Coloradans. No one in California, Oregon, Nevada or Utah has ever been dispatched by a twister.

Some locales seem more prone to tornado strikes than others. Western "tornado alleys" are, like everything else related to tornadoes, smaller than their midwestern counterparts. In the Midwest, tornado alleys are measured in states; western tornado alleys extend across counties. Some of the West's favorite tornado hangouts are the Columbia River Basin from Portland to Spokane, the Los Angeles Basin, the Gila Desert around Phoenix, the Snake River Plains of Idaho and several swaths across the High Plains.

All of these tornado alleys are wide, flat areas where storms can spin undisturbed by mountains. Over the High Plains, the swaths may follow zones of eddies downwind from mountains and ridges. (There's more about this in the chapter on whirlwinds and washoe zephyrs.) Most of the alleys are in populated farm country or even cities, where residents notice tornadoes. Many twisters probably go undetected as they tumble sagebrush across some uninhabited desert.

As the population of the West increases, so does the tornado count. Consider the words of Lorin Blodget in his 1857 *Climatology of the United States*: "The frequency and distribution of these tornadoes is a subject of practical interest. There are none on the Great Plains so far as known. They are most numerous in the Mississippi Valley." But that was back when the Great Plains had few people to notice tornadoes; we now know that the area breeds fully half the tornadoes in the United States, and that it has more tornadoes per square mile than any other place in the world. As tornado reports continue to rise in the West, we can expect to hear a lot of discussion about changing storm patterns. It may really be changing people patterns.

THE BIG STORMS

December 27, 1866—A tornado in Nevada County, California, may be the first ever reported in the West.

July 29, 1883—The snail's pace of geology speeded up briefly, as a flash flood dug a 50-foot-deep gorge along Kanab Creek, Utah.

June 14, 1903—A cloudburst unleashed a flood that washed away Heppner, Oregon, with a loss of 200 lives.

August 7, 1904—Eighty-nine passengers drowned as a train trestle, weakened by a flash flood, collapsed near Pueblo, Colorado.

August 10, 1924—The West's most lethal tornado ever struck Thurman, in northeastern Colorado, killing 10 occupants of a single house.

June 19, 1938—Another train disaster caused by a flood-weakened trestle. The "Olympian," bound from Chicago to Tacoma, plowed into Custer Creek, near Miles City, Montana, with a loss of 49 lives.

December 2, 1970—A "white tornado" swept across Timpanogos Divide, Utah, snapping trees and lifting snow 1,000 feet up into its funnel.

April 5, 1972—A quarter-mile-wide tornado that touched down in Portland, Oregon, crossed the Columbia River to Vancouver, Washington, where it devastated a shopping center and killed six people.

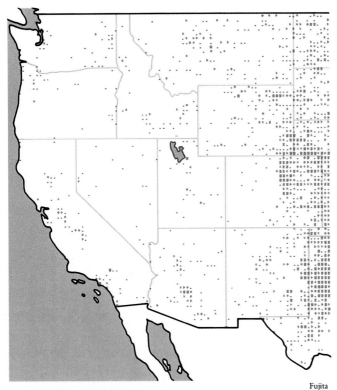

TORNADOES IN THE WEST—This map, produced by Theodore Fujita of the University of Chicago, locates every tornado known to have touched down in the West over a 66-year period. Note that tornadoes prefer open, flat terrain, such as eastern Colorado and California's Central Valley. However, there are also more tornadoes near major population centers—do they really like cities, or is there simply better reporting in urban areas?

Fujita

June 9, 1972—A stationary thunderstorm dumped up to 15 inches of rain on the Black Hills of South Dakota, sending torrents of water into Rapid City and other towns. It was South Dakota's worst disaster, with 236 people killed, 2,900 injured and nearly 1,000 homes destroyed.

August 7, 1975—A Continental Airlines 727 jet encountered a downburst as it took off from Denver. The plane fell back to the runway and skidded into a field, injuring 15 passengers.

July 31, 1976—On the eve of its 100th anniversary of statehood, it was Colorado's turn to suffer its worst natural disaster. A stationary thunderstorm, very similar to the Rapid City storm, flooded the Big Thompson River. Of the estimated 3,000 people (many of them vacationers) in the canyon at the time, 139 died in the flood.

May 18, 1977—A huge, cone-shaped tornado cut a 42-mile-long path from Oklahoma into extreme southeastern Colorado. Most of its track was over open country but the storm struck several farms, tossing trucks, farm machinery and cattle up to one-half-mile away. Fortunately, no people were injured. Damage patterns indicated peak winds exceeding 260 m.p.h., making this the most powerful tornado ever known to have struck a western state.

March 4–5, 1978—Six tornadoes in two days touched down around the north end of Sacramento Valley, while waterspouts churned waters off Santa Monica. Damage was limited to several barns and chicken coops.

July 16, 1979—A tornado cut across the northern part of Cheyenne, Wyoming, destroying planes and hangars at the airport and more than 150 houses and trailers. Thirty minutes later a much weaker twister touched down in eastern Cheyenne.

June 3, 1981—Three tornadoes touched down from a powerful thunderstorm that swept northward across Denver. The second tornado destroyed 87 homes in the northern suburb of Thornton, but thanks to timely warnings there were no deaths.

November 9, 1982—A Pacific storm spawned seven tornadoes in the Los Angeles area. One twister cut a 10-mile-long path along the Los Angeles River through Long Beach, another blew a catamaran 100 feet into the air at Malibu, but in general the damage was minor.

March 1, 1983—Another Los Angeles tornado came within a mile of the Civic Center, damaging 50 homes and injuring 30 residents. Two more twisters in September 1983 brought the 12-month total in the Los Angeles Basin to 11!

August 17, 1984—A short-lived tornado was sighted above timberline on Colorado's Longs Peak. At 11,400 feet elevation, it may be the highest ground ever crossed by a twister.

July 31, 1987—A massive, 4,000-foot-wide funnel sliced through subdivisions and industrial parks on the east side of Edmonton, Alberta, leaving 27 dead, hundreds injured, and $250 million in damage. It was the deadliest tornado ever to touch down in western North America.

July 11, 1990—The most devastating hailstorm in U.S. history hammered a 10-mile-wide, 130-mile-long swath along the Colorado foothills from Estes park to Colorado Springs. Around Denver half-pound hailstones mashed roofs and windshields on more than 100,000 cars (including mine!), punctured thousands of houses, and bruised dozens of riders trapped on stalled rides at an amusement park. Total losses exceeded $600 million.

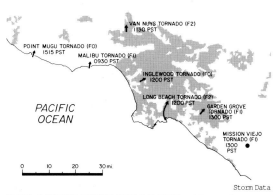

Storm Data

LOS ANGELES TORNADOES—Tornadoes struck the Los Angeles area on November 9, 1982. The path of the Long Beach tornado is shown below.

Storm Data

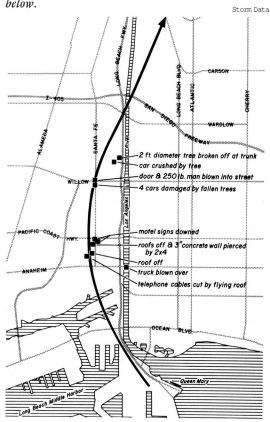

HURRICANES

To the Indians of the Caribbean they were known as *Huracan*—the "Storm Devil." The Indians of India call them *cyclones*, a word of Greek heritage passed on by the British. In Chinese the name is *typhoon*, or "great wind." And Australians, for some peculiar reason, have been heard to call them "Willy Willies." We call them hurricanes, and in any language they are the greatest storm on Earth.

THE GREATEST STORM ON EARTH

Frightful stories of hurricanes' power and destruction abound in the annals of history. Columbus lost two ships to a hurricane during his second voyage to the New World in 1494. Four and a half centuries later, on December 18, 1944, the United States Third Fleet was attacked by a small but intense Pacific typhoon named Cobra. In a few hours, the tempest had claimed three destroyers, 146 aircraft and 790 sailors. It was one of the greatest disasters inflicted on the American Navy during the final year of World War II. Many a survivor of a hurricane at sea shares Shakespeare's account of Casca's stormy premonition of Caesar's death: "I have seen the ambitious ocean swell, and rage, and foam, to be exalted with the threat'ning clouds. . ."

It is when hurricanes leave the sea and come ashore that they unleash their greatest wrath on humanity. A hurricane from the Bay of Bengal drowned 300,000 people in 1737, and as recently as 1970 another hurricane killed half a million citizens of the then-soon-to-be nation of Bangladesh. The greatest natural disaster ever to befall the United States was the devastation of Galveston, Texas, by a hurricane in 1900. One-sixth of the city's 40,000 residents perished when the storm pushed the Gulf of Mexico several miles inland. The power of storm waves striking shore is so great that seismographs—designed to measure earthquakes—in Alaska detected the rumble of crashing seas from the New England hurricane of 1938. Hurricanes are truly earth-shaking storms!

72

Galveston, New England and Bangladesh are a long way from the West, and it might seem that hurricanes are not part of the West's varied weather. It is true that no full-force hurricane has ever directly struck any western state. But tropical storms with winds just below hurricane force (74 m.p.h.) have on several occasions wreaked considerable damage to western towns and cities. Nearly every year the West's weather is affected by dying remnants of hurricanes whose lives began in such distant places as Micronesia and West Africa. Before we can appreciate the impact of hurricanes on western weather, we must first understand what a hurricane is, and what makes them so different from the multitude of other storms that darken western skies.

Hurricanes are creatures of the sea. Their breeding grounds are the warm oceans of the tropics, and they grow as long as their travels continue to take them over the tepid seas. Unlike mid-latitude cyclones, which draw their power from the temperature differences across frontal boundaries, hurricanes thrive on undiluted warmth. Warm water feeds tremendous volumes of water vapor into the tropical atmosphere. This moisture-laden air is the fuel that drives hurricanes; without it, the great storms weaken and die. Like many other sea creatures, hurricanes soon expire when they leave the sea and go ashore.

The hurricane season begins in late summer, when the tropical (between 10 and 30 degrees north latitude) Atlantic and Pacific oceans warm to their highest temperatures of the year. Several months of nearly overhead sun has heated the water to 85 or 90 degrees; at these temperatures water readily evaporates into the overlying atmosphere. The dew point of the tropical atmosphere may rise above 80 degrees. Such oppressive humidity is rarely experienced anywhere in the western United States. The huge volume of water vapor in tropical air contains a tremendous amount of energy in the form of latent heat—heat stored in molecules of water vapor when they evaporated from the ocean surface. When vapor condenses back into liquid water, the latent heat is released back to the air.

The air overlying just one square foot of the tropics may hold enough energy as latent heat to drive an automobile two miles. Multiply this by tropical oceans that cover tens of millions of square miles! In the tropics, latent heat is normally released back into the atmosphere by the same process of convection that gets thunderstorms going. The moist air of the tropics needs to rise only 1,000 or 2,000 feet before condensation creates cumulus clouds. The release of latent heat warms the air, and the clouds grow into towering cumulus and then go on to become cumulonimbus clouds. The cycle of evaporation, rising streams of air and condensation is the driving force behind hurricanes. Like the ignition of gasoline vapor inside the cylinders of an automobile engine, it is the way hurricanes use their fuel to create motion.

For months, from late summer well into the fall, vast reaches of the tropical atmosphere are ripe for hurricanes to form. But hurricanes are relatively rare, and every year only a few dozen are born worldwide. It takes more than hot, muggy air to give birth to a hurricane—something needs to lift the air to its condensation level. Even this lifting is not enough, though, since cumulus clouds are commonplace in the tropics. Although many cumulus clouds grow into thunderstorms and even huge masses of thunderstorms hundreds of miles across, these thunderstorm masses usually lack the intense, swirling winds of a hurricane. Early in the life of a hurricane, conditions must be just right or the great storm will never develop. Meteorologist William Gray of Colorado State University has examined hundreds of cases of hurricane formation worldwide and found that hurricanes need some sort of "seed" disturbance—a slight swirl in the atmosphere—to get things going

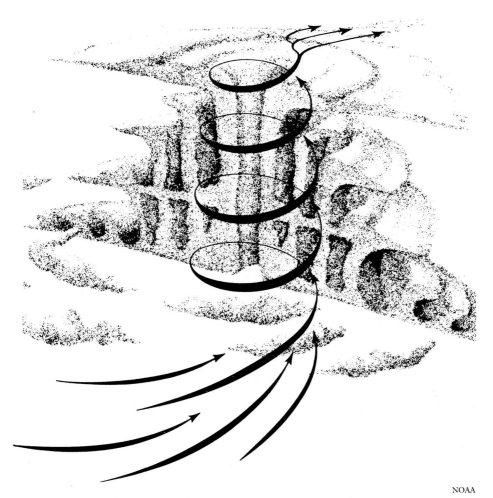

NOAA

CROSS SECTION OF A HURRICANE—Moist air spirals into the storm at low levels, then corkscrews upward near the center. The rising currents near the center form enormous, rainy clouds that surround that bizarre feature of all hurricanes—the clear eye.

on the right track. He also found that if winds near the thunderstorm tops were too strong or blew in a direction different from the low-level winds, the developing thunderstorms would be torn apart as they reached the upper atmosphere. Hurricanes indeed have difficult childhoods.

Sometimes a twist in the trade winds or a weakening cold front from the north provides the initial impetus. The "seed" may be an exceptionally intense mass of tropical thunderstorms. Whatever its origins, the incipient hurricane starts as a slowly rotating collection of thunderstorms. Warm, moist air from outside the disturbance spirals slowly in, replacing the air that rose upward

HURRICANE NORMAN—Bands of thunderstorms spiral in toward the center of Hurricane Norman, located off the tip of Baja California on September 2, 1978. At the time of this satellite image, winds of 140 m.p.h. churned the ocean near the storm's 40-mile-wide eye.

in the thunderstorms. The Coriolis effect assures that this spiral will, in the Northern Hemisphere, turn counterclockwise. Like an upside-down version of a draining bathtub, the inward flow spirals faster and rises as it approaches the center. On their way to the storm center, the winds evaporate more moisture from the ocean surface; this moisture then drives the convection at the core of the spiral. The cycle continues to accelerate until the swirling winds reach gale force (39 m.p.h.), and the tempest becomes a "tropical storm." When the winds reach 74 m.p.h., their circular motion becomes so strong that a calm spot—the eye—forms at the center. The storm is now a hurricane.

A fully grown hurricane is a marvel of natural engineering. In a continuous, mechanical cycle it extracts latent heat energy from the ocean, releases the energy in the towering thunderclouds surrounding the eye, and exhausts the spent air—with much of its moisture gone—out the top of the storm. The amount of energy involved in a hurricane is almost beyond comprehension, and renders humanity's most awesome efforts—nuclear weapons—feeble by comparison. The heat energy contained in the combined nuclear arsenals of the United States and the Soviet Union would not keep a large hurricane going for one day. Many will recall photographs of the 1946 underwater nuclear explosion in the Pacific, which tested the effects of the blast on captured Japanese warships. That detonation heaved 10 million tons of water into the air. A good-sized hurricane can evaporate that much water from the ocean in 40 seconds.

These tremendous energy releases give hurricanes their purpose in life. In the global atmospheric ecosystem, the prime function of all storms is to equalize the great heat imbalance that builds up between the equator and the poles. At higher latitudes, it is the familiar mid-latitude cyclone that does the job. In the tropics, hurricanes are particularly efficient at moving heat. They extract latent heat from the oceans, convert it to sensible heat and lift it to the upper atmosphere, from where the heat is blown poleward.

Most hurricanes wither and die when they encounter cold water. Cold water evaporates less readily than warm water and, without a continuous supply of water vapor, hurricanes soon "run out of steam." Some hurricanes are deprived of their needed vapor supply when they pass over land, but the effect is the same—oblivion. When hurricanes die, however, there is always something left over. Remnants may take the form of a mass of moisture-laden air, a slight residual swirl in the upper atmosphere or both.

HURRICANE ALLEY

Some parts of the tropics are more amenable to hurricanes than others. Gray identifies seven regions around the globe—the North and South Indian oceans, the seas northwest of Australia, the South Pacific, the western and eastern North Pacific oceans and the western North Atlantic—where hurricanes and tropical storms can develop. None has ever formed over the cool waters of the tropical South Atlantic or southeastern Pacific. The western North Pacific Ocean is the most active region, with one-third of the globe's average of 79 tropical storms per year. Surprisingly, the second-busiest hurricane alley is the eastern North Pacific, averaging 13 storms per year. Most of these storms form off the west coast of Mexico and move west toward Hawaii (but few ever reach the islands). This reach of the Pacific has few islands and sparsely traveled shipping lanes, so many hurricanes went undetected in the days before weather satellites. It wasn't until the 1960s, with the advent of regular satellite coverage of the world's weather, that the importance of this hurricane alley became fully appreciated.

Mexicans living along the west coast from Acapulco to Guaymas and southern Baja California have long appreciated these storms. Once every few years a hurricane sweeps northwestward along

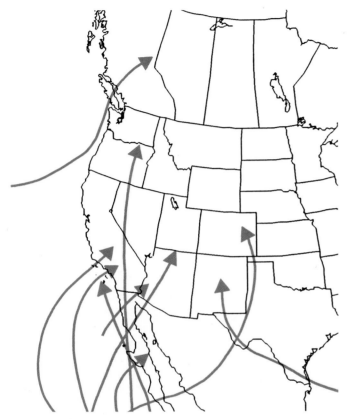

HURRICANE ALLEY—The tracks of 10 hurricanes (or tropical storms) that have affected the West are shown on this map. Most are "Cordonazo" type storms that move up the west coast of Mexico, while a few have stormed up the Rio Grande Valley from the Gulf of Mexico. The remains of one western Pacific typhoon—Freda—crossed the ocean to become the Columbus Day storm in 1962.

the coast before turning inland, wreaking havoc among fishing fleets and their home villages. Peak season for these coastal storms is September and October, centering around the Feast of Saint Francis on October 4. This has earned the storms the local name *el Cordonazo de San Francisco*— "The Lash of Saint Francis." The southwestern United States has also suffered from el Cordonazo, and indeed most, but not all, of the tropical storms to strike the western states are of Mexican birth. Hurricane remnants from the Gulf of Mexico sometimes reach into the western states, and on one occasion the remains of a western Pacific typhoon inflicted heavy damage to the Pacific Northwest.

From the moment a hurricane is born, its movement is guided by prevailing winds. While still in the tropics (10 to 30 degrees latitude), hurricanes usually move westward with the easterly trade winds. Once north of the 30th parallel, the storms encounter the prevailing westerlies of higher latitudes and tend to turn eastward. The typical path of a well-behaved hurricane is a sweeping arc, at first heading west, then turning north, and finally heading northeast toward the cooler waters of the northern oceans. These guiding winds, called "steering currents," are not always well-behaved, however. Hurricanes have been known to backtrack, loop and even stand still for days on end. Sometimes the steering currents bring tropical storms into the western states.

Take el Cordonazo, for example. Steering

NSIDC, University of Colorado

HURRICANE NORMAN APPROACHES CALIFORNIA—Clouds and moisture from a weakening Hurricane Norman stream into central California on September 4, 1978—two days after the earlier photo of Norman. The next day Norman dissipated off the Southern California coast, but its moisture brought heavy rains and floods to the deserts. The Salton Sea and many smaller geographical features are easy to spot in this remarkable satellite image.

currents carry most of the storms northwest to cooler waters. A few turn north and move inland over Mexico. Fewer still move as far north as southern California, where chilly coastal waters quickly sap their vigor. Once in a blue moon tropical storms race north into the United States before losing their strength, and the normally dry early autumn in southern California or Arizona suddenly turns windy and wet.

No full-force hurricane has ever been known to directly strike the western states. However, twice in this century el Cordonazo storms of near-hurricane intensity damaged cities of the southwestern United States. The first was the tropical storm of September 1939, which stormed straight into Los Angeles with gales and torrential rains, claiming 45 lives. In 1976 Hurricane Kathleen, which had just been downgraded to tropical storm status, lashed Yuma, Arizona, with 76 m.p.h. gusts. Remnants of this remarkable storm were tracked as far north as Oregon and Idaho. Every few years the fading remains of a tropical storm drift into the southwestern states, dumping heavy rains onto normally dry deserts and mountains. And nearly every year, the counterclockwise circulation around one or two hurricanes off the West Coast drives moist air into the Southwest, triggering tropical downpours that have occasionally developed into floods.

WESTERN HURRICANES

September 25, 1939—A long-lived Cordonazo that formed south of El Salvador on the 14th came ashore over Los Angeles. Gales exceeding 50 m.p.h. damaged boats, buildings, and power and telephone lines along the coast to the tune of $1.5 million. Forty-five lives were lost. Torrential rains—5.42 inches in Los Angeles and 13 inches on Mount Wilson—ruined half a million dollars worth of crops. This was the third—and strongest—Cordonazo to ruffle southern California in 18 years.

September 1, 1942—The remains of a powerful Gulf of Mexico hurricane crossed Texas and stalled over eastern New Mexico. Rains in excess of six inches flooded the Pecos River near Roswell.

August 26–28, 1951—This Cordonazo brought three days of steady rain to central Arizona, totaling eight inches in some places.

July 16–20, 1954—The moisture from a dying cyclone over the Gulf of California triggered five days of heavy thunderstorms and local flooding over extreme southern Arizona.

October 12, 1962—The "Columbus Day storm," with hurricane-force winds and heavy rains, ripped Washington and Oregon at a cost of 56 lives and $250 million in damages. Wind gusts of 170 m.p.h. were reported along the Oregon Coast. The storm originated as Typhoon Freda, born nine days earlier near Wake Island in the western Pacific. Freda merged with a cold front over the North Pacific to become an intense mid-latitude cyclone, which raced eastward into the Pacific Northwest.

September 18, 1963—Short-lived Tropical Storm Katherine dissipated over Yuma, where it dumped a record 2.42 inches of rain—nearly the normal annual total—in less than three hours. Flood damage to property and lettuce crops was considerable.

September 2, 1967—Yuma was struck again, this time by ex-Hurricane Katrina, which had followed a path up the Gulf of California. Two inches of rain and 49 m.p.h. winds hit the desert city.

September 4–6, 1970—Although Tropical Storm Norma remained hundreds of miles to the south, her circulation pumped enough moist air over the Southwest to trigger the deadliest flooding in Arizona's history. Flash flooding in canyons and streams across Arizona and neighboring sections of Utah, Colorado and New Mexico drowned 25 people and washed out numerous

Wes Luchau

TOPPLING TOWER—Photographer Wes Luchau of Western Oregon State College caught the 1962 "Columbus Day storm" in the act of toppling the tower on Campbell Hall on the college's campus in Monmouth, Oregon.

roads and bridges. Thunderstorms also brought small tornadoes to Scottsdale, Arizona, and Cortez, Colorado.

September 29, 1971—The remnants of Hurricane Olivia brought generally welcome rains to Arizona and New Mexico. Olivia originated as Hurricane Irene, born 800 miles east of Barbados in the Atlantic, and was renamed when the storm crossed Central America and entered the Pacific. This long-lived storm endured 19 days.

October 6, 1972—The center of Tropical Storm Joanne passed northward across Arizona with 45 m.p.h. winds and heavy thunderstorms. Flood damage to roads and the cotton crop totaled $5 million.

September 10, 1976—Hurricane Kathleen,

80

weakening but still strong, came ashore south of Ensenada, Mexico, and raced north over Calexico and Death Valley, California, and into western Nevada. Up to 10 inches of rain flooded the deserts of southern California, unleashing a 15-foot wall of water that roared into the small town of Ocotillo, killing six. Seventy-six-m.p.h. gusts inflicted $2 million in damages to Yuma. The next day, rains and gusty winds moved as far north as Oregon, Idaho and Montana, flooding basements in Boise. Before it was all over, this remarkable storm had dumped heavy rains in every state in the West.

August 14–17, 1977—While tracking northward along the Baja California coast, Hurricane Doreen brought flash floods to Imperial Valley, California, and southeastern Arizona. Several irrigation canals were breached, inundating crops and the town of Niland, California. In Yuma, Arizona, three-inch rainfall was accompanied by 56 m.p.h. winds. Los Angeles, where August is normally rainless, received 2.47 inches from the storm.

September 6, 1978—Ex-Hurricane Norman dissipated near San Clemente Island, off the southern California coast, after causing local flooding in the deserts. This marked the third year in a row that this area sustained flooding.

August 14, 1980—One of the most intense Atlantic storms in history, Hurricane Allen, came ashore in Texas and moved up the Rio Grande valley toward New Mexico. Five inches of rain flooded streets and homes in the northeast section of Albuquerque. Allen originated as a disturbance off the west coast of Africa and tracked 5,000 miles to dump its rain on Albuquerque. At one point its winds reached 190 m.p.h.

September 26, 1982—Remains of Hurricane Olivia came ashore near Santa Barbara, California. Heavy rain soaked half of California's raisin crop, and agricultural losses totaled more than $300 million. Moisture streamed into Utah,

flooding parts of Salt Lake City.

September 30–October 3, 1983—Although its center remained well to the south, Tropical Storm Octave pumped moist air into the Southwest, triggering Arizona's most destructive flood in history. Thirteen people died in the deluge. The same moist air mass bred two tornadoes that struck the Los Angeles area, but damage was comparatively minor.

September 10, 1984—Moisture from Hurricane Marie, 300 miles south of Los Angeles, triggered street flooding in Las Vegas and Phoenix. Five people drowned when their pickup truck was washed away.

October 1–4, 1984—Remains of Hurricane Polo came ashore north of Mazatlan, Mexico, and tracked northward across New Mexico into Colorado. Heavy rains caused minor flooding along the east slopes of the Rockies. Above timberline a foot of snow fell at Echo Lake, west of Denver.

YUMA—THE WEST'S "HURRICANE ALLEY"

It may seem ironic that Yuma, Arizona, the driest city in the United States, has probably been affected by tropical storms more often than any town "west of the Pecos." But a sizeable portion of Yuma's meager rainfall does result from tropical storms passing nearby or from moisture brought in from more distant storms. Yuma's annual rainfall averages just under three inches, and in one year (1956) it was just a tenth of that. It is not surprising that a large portion of the yearly total can arrive in one storm. When that happens, the storm is likely to be a hurricane remnant. Of the past 25 years, four—1963, 1967, 1972 and 1977—saw more than half the year's total rain fall from a dying hurricane. Hurricane Katherine dropped half of 1963's rain in just over one hour!

The other attendant of tropical storms, gale-force winds, visited Yuma in 1967, 1972, 1976 and 1977. That's about once every five years, nearly as often as many cities in Florida can expect tropical gales. Why does Yuma experience more tropical storms than coastal cities like Los Angeles or San Diego? It's the water. Although a desert city, Yuma lies just 70 miles north of the tepid waters of the Gulf of California, which can simmer at 90 degrees in the summer. Tropical storms approaching Yuma can therefore get a last quick fix of water vapor before reaching land. On the other hand, any storm headed for southern California is sapped by chilly coastal water. As a result, Yuma's 43,000 inhabitants can not only claim to live in the hottest, driest and sunniest city in the United States, they can also claim to live in the heart of western America's hurricane alley!

LOOKING ELLA IN THE EYE

All sorts of superlatives can be applied to describe the power of a hurricane, but the surest way to appreciate the tempest is to encounter it personally. During World War II the Navy experimented with flights into hurricanes to pinpoint their positions. The flights began in earnest following the Third Fleet disaster in 1944, and have since become routine. Few experiences impart a person with a greater sense of awe at nature's majestic might than a flight into the eye of a hurricane. In 1978 I had the good fortune to fly on a National Oceanic and Atmospheric Administration (NOAA) flight into Hurricane Ella off Cape Hatteras.

The NOAA Orion—a military version of the Electra, a four-engine propjet—took off from Miami as dawn was breaking. An hour after sunrise, the first squalls on the storm's fringe appeared on the horizon. Below, the ocean surface was an undulating pattern of long, rolling swells fleeing from Ella. A freighter wisely fled with the swells.

Minutes later the hurricane-hunting Orion began its bumpy journey through these outer fringes of Hurricane Ella. Viewed through the Orion's panoramic bubble windows, the squalls appeared as long bands of billowing cumulus clouds, curving in toward the storm's center, now 100 miles away. The many scientific instruments on board the $8 million flying laboratory went into action—recording temperature, humidity, wind and even the size of the droplets in the clouds. Scientists strapped with seat belts in front of consoles pushed buttons and flicked switches; charts and numbers appeared on television screens.

On radar, the entire hurricane was in clear view. Its spiral pattern could have been a telescopic view of some distant galaxy, except for its closeness and the hole in its center. The Orion drove on toward Ella's core, sweeping in and out of the spiral rainbands; 700 feet below, turquoise streamers of underwater foam—churned by the furious winds—stretched to the horizon.

Several severe jolts announced the plane's entry into the ring of eight-mile-high rain clouds surrounding the eye of the storm. Winds of 100 miles an hour—momentarily gusting to 150 m.p.h.—whipped an almost-solid sheet of foam across the surface of the sea. It seemed as if the waves were being flattened by the very winds that had created them.

Suddenly—within seconds—the jolts ceased. The Orion had entered the calm eye of Ella. Below, only a few whitecaps marred the blue waters of the sea. Above, the sun occasionally peeked through thin clouds. As the Orion darted in and out of puffy gray clouds, hurricane hunters on board caught occasional glimpses of the eyewall itself—an awesome amphitheater of massed clouds, dimly visible through the mists.

The respite in the 10-mile-wide eye lasted only two minutes. The plane plowed back into the violent winds of the eyewall and for the next half-hour bounced its way back to the relatively calm fringes of Ella. When the Orion was safely on its way home to Miami, the old-timers on board began telling their tales of lost wingtips, ruptured gas lines and disabled navigational equipment.

A TEMPEST IN A TEACUP

It's easy to demonstrate how hurricanes work by doing a little experiment right in your kitchen. Make a cup of tea using loose tea or a torn-open tea bag. This gives you plenty of tiny leaves on the bottom of the cup. Stir the tea, in a counterclockwise direction if you like, with a spoon. If you stir it hard enough, a dip forms in the center of the cup. Satellite photos of hurricanes show that the clouds around the eye have the same shape as the tea surface around this dip. Now look at the tea leaves. They are piling up in the middle of the cup, thanks to a current of tea moving toward the center along the bottom, just like the low-level flow into a hurricane. The lighter tea leaves swirl up around the center toward the surface, mimicking the rising air currents near the eye of a hurricane. Near the surface, the tea leaves spiral away from the "eye" and sink back down near the edge of the cup. These currents of tea are virtually identical to the motions of air in a hurricane, with one important difference. Hurricanes keep themselves going by releasing latent heat in their centers. Tea leaves contain no latent heat, so the energy must come from the stirs of the spoon.

WHIRLWINDS & WASHOE ZEPHYRS

This was all we saw that day, for it was two o'clock, now, and according to custom the daily 'Washoe Zephyr' set in; a soaring dustdrift about the size of the United States set up edgewise came with it, and the capital of the Nevada Territory disappeared from view. Still, there were sights to be seen which were not wholly uninteresting to new comers; for the vast dust cloud was thickly freckled with things strange to the upper air—things living and dead, that flitted hither and thither, going and coming, appearing and disappearing among the rolling billows of dust—hats, chickens and parasols sailing the remote heavens; blankets, tin signs, sage-brush and shingles a shade lower; door-mats and buffalo robes lower still; shovels and coal scuttles on the next grade; glass doors, cats, and little children on the next; disrupted lumber yards, light buggies and wheelbarrows on the next, and down only thirty or forty feet above the ground was a scurrying storm of emigrating roofs and vacant lots.

—Mark Twain, *Roughing It*

Mark Twain first encountered the fantastic variety of western climates during a stagecoach trip from Missouri to Nevada in 1862. In *Roughing It*, he recounts the approach of a "Washoe Zephyr" during his first day in Carson City, Nevada. Along with chinooks, lee waves, Santa Ana winds, sea breezes, valley winds and those little whirlwinds called dust devils, Washoe Zephyrs are one of the multitude of local winds that color western weather. In a sense, there are as many local winds as there are hills and valleys, plus a few more. Many of these winds have a lot in common, though, and fortunately there's no need to describe each one individually. Instead, we'll look at the main types of local winds that occur and point out some of their local characters.

CHINOOKS

Once you get away from the coast, there's hardly a place in the West that doesn't get a chinook once in a while. The name comes from the

Chinook Indians of coastal Oregon, who called the warming southwest winds off the ocean "snow-eater." Now the name is usually given to warm westerly winds that eat the snow in the valleys of the Sierras, Cascades and Rockies.

Snow-eaters can be a welcome treat in the winter. Chinooks can lift temperatures 20 to 80 degrees in a few hours and melt half a foot of snow in an afternoon. A chinook at Kipp, Montana, reportedly ate 30 inches of snow in half a day. But chinooks can also tear roofs from houses and blow trains from their tracks. They are truly mixed blessings. Whether you like them or not, however, they are as much a part of the West as cowboys and rattlesnakes.

Most of the mountain ranges of the West run roughly north-south, while, most of the time, winds blow from west to east. This is why chinooks are so common. When a west wind meets a long range of mountains blocking its flow, it has no choice but to go up and over. When conditions are right, which they often are, the air continues to go up and down after it has passed the mountain ridge. The best way to visualize this is to watch what happens in a shallow stream when the water runs over a rock. The water goes up and over, and downstream there's a series of ripples. Air does the same thing. As far as the laws of physics are concerned, air is a fluid, just like water. Air may be lighter, and water wetter, but they're both still fluids.

These ripples of air are known as "mountain waves" or, by virtue of their position downstream from mountains, as "lee waves." Just like the creek waves, they stand still while the air flows through them. And also like the creek, the air flows fastest at the bottom—or trough—of the waves, especially in the first trough. These features are very important for the kind of weather these waves can bring.

Mountain waves are a lot bigger than anything you'll see in a creek. Ripples can be 10 miles apart, and the air going through them may rise and fall a mile or more. Mountain waves can also extend upward through the atmosphere—they've been measured as high as 15 miles above the ground, where the air is only 5 percent as dense as at sea level. Glider pilots love them because they

MOUNTAIN WAVES—Like ripples in a creek, waves form downstream from mountain ranges.

U.S. Forest Service

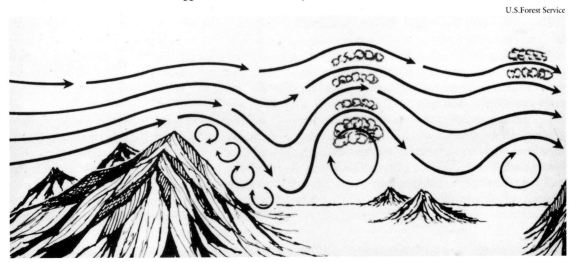

85

often provide a reliable, stationary and strong upward current of air—great for hitching a soaring ride! Many gliding records for altitude and duration came from pilots successfully "catching a wave," and their exploits have contributed greatly to our knowledge about mountain waves.

If chinook means "snow-eater," what makes these winds so warm? One common explanation says that as air goes up one side of the mountain, it cools and its water vapor condenses, releasing latent heat into the air. Since the air has gained a certain amount of latent heat, it's warmer when it comes back down the other side of the mountain. If air blowing off the Pacific has a lot of moisture, it can warm as much as 10 degrees as it crosses the Sierras and Cascades. Over most of the West the air is simply too dry, and the release of latent heat warms the air only a degree or so. Usually, though, there is enough moisture for "cap clouds" or "crest clouds" to form right on the mountain ridge.

The main reason chinooks are warm is very simple—their air comes from a relatively warm place! The westerly winds that bring chinooks also bring mild air off the Pacific Ocean. The air further warms by compression as it blows downhill in a mountain wave. Inland, the winter air masses are usually a bit colder than this temperate Pacific air, and are sometimes much, much colder. Some spectacular temperature rises have happened when chinook winds replaced arctic air masses. In the next chapter you'll see exactly how rapidly a chinook can warm things up.

There's no law that says chinooks *must* be warm. If the westerlies drive a cold air mass into the West, the chinooks are cold. When the new air is colder than the air mass that's being blown away, the wind is properly called a *bora*, originally a Yugoslavian word for the cold winds that pour through their mountains. Usually, however, chinook winds are warm.

The welcome warming a chinook brings can exact a high price at times. Remember that the air flow through a mountain wave is fastest at the bottom of the first trough east of the mountains. These "downslope winds" often exceed 100 miles an hour. Since the mountains don't move, the troughs don't either, and their powerful winds ventilate the same place year after year. The highest winds nearly always occur in a narrow band within 10 or 20 miles of the foothills.

The Ute Indians of Colorado knew this and would not set their tepees at certain places just east of the Rockies. The white man didn't know this, or ignored it if he did, and in 1859 settled at the mouth of Colorado's Boulder Canyon. Every few years since then, the town (now a city) of Boulder has been raked by chinook winds strong enough to raise roofs and down power lines. A windstorm in 1969 did a million dollars damage, and just three years later another tripled the damage toll. A 1982 chinook cost the insurance industry $17 million.

The same story can be told in Livingston, Montana; Sheridan, Wyoming; and Bishop, California; as well as in Carson City, Nevada—home of the Washoe Zephyr. Salt Lake City can be whisked by chinook winds, even though its mountain range, the Wasatch, is to the *east*. This happens when a strong low-pressure system moves south of Utah, with high pressure to the north. The winds then come from the east, and the mountain wave develops west of the Wasatch—right over Salt Lake City. Although rare, these easterly chinooks have been every bit as strong as their more common westerly cousins.

Besides melting snow and wrecking cities, chinooks also make for the most spectacular sunsets on earth. As the air streaming through a mountain wave rises from trough to crest, its moisture condenses into little clouds that sit atop the waves. Because of their smooth, rounded, lens-shaped appearance, these clouds have earned the name "lenticular clouds," although others prefer to call them "grindstone" clouds. These weird

Jose Meitin

CHINOOK WINDS—Power line poles in Boulder, Colorado, were toppled like falling dominoes by the chinook windstorm of January 17, 1982.

clouds will stand still for hours, even though the wind is ripping through them at 100 miles an hour or more. Look at one with binoculars; you'll see tiny shreds of cloud form on the upwind edge, blow through the lenticular and dissolve on the downwind edge.

The brilliant sunsets come just before the sun dips behind the mountains to the west. The descending air flow between the mountains to the first wave trough leaves a wide clear slot, allowing the sun to illuminate the wall of lenticular clouds that rises to the east. Many a winter day comes to a crimson close in chinook country.

SANTA ANA WINDS

Another wind that charges a high price for the good it does is the Santa Ana wind of southern California. These hot, blustery gales are very similar to chinooks, but since they whip through the palms of Los Angeles, they never get a chance to eat much snow. The meteorological setting calls for high pressure to settle over the Great Basin, with lower pressure to the south. With the help of upper-level winds out of the north, the winds stream up and over the San Gabriel, San Bernardino, and Santa Ana mountains, and descend into the vast basin occupied by Los Angeles and environs.

These northerly or northeasterly winds can blow any time of year, but they prefer late fall and winter. Like chinooks, they have unroofed homes and downed trees and power lines. Their good side is that they blow the smog far out to sea.

By the time the winds reach Los Angeles, they have dropped several thousand feet from the Great Basin. Thanks to compressional heating, the temperature has gone up and the humidity down, which, combined with the gusty winds, makes for serious fire danger. Nearly every year, especially in the fall when foliage is dry, fires break out during spells of Santa Ana winds. At times these fires have turned into major disasters, consuming hundreds of homes and thousands of acres of forest.

Other places along the coast can also get these desiccating gales, although not nearly as often as Los Angeles. This is probably because Los Angeles gets struck when the wind is from the north, while most of the coast needs a rarer east wind. But San Francisco and San Diego have seen their own versions of the Santa Ana, and sometimes with the same results—fire! Probably the most dramatic was the weak, but still dry, easterly wind that swept fire across San Francisco on April 18, 1906—the day of the great earthquake.

LEE EDDIES

No, these aren't the names of a couple of good ol' boys. These are eddies, or swirls, of air that form in the lee of mountains. They're not mountain waves, or even lee waves. Picture a stream again, this time with a bigger rock or shallower water, so water flows *around* the rock, not over it. See how the water whirls around the backside of the rock? That's a lee eddy. Kayakers know these well and take great pains to avoid the greater pain of getting caught in one.

Lee eddies come in many sizes. The largest are the cyclones that redevelop east of the Rockies, described in the chapter on fronts, jets and cyclones. Next down in size are the county-sized swirls that develop downstream from county-sized mountain ranges. Meteorologists at the University of Washington have discovered that little cyclones, 20 miles or so across, develop east of the Olympic Mountains when the wind is from the west or southwest. These little cyclones are no laughing matter—one of them took down the Hood Canal Bridge.

Researchers at the National Oceanic and Atmospheric Administration labs in Boulder, Colorado, have found that a similar but weaker cyclone often develops north and east of Denver. This "Denver cyclone" happens when southwesterly winds curl around a low ridge south of town.

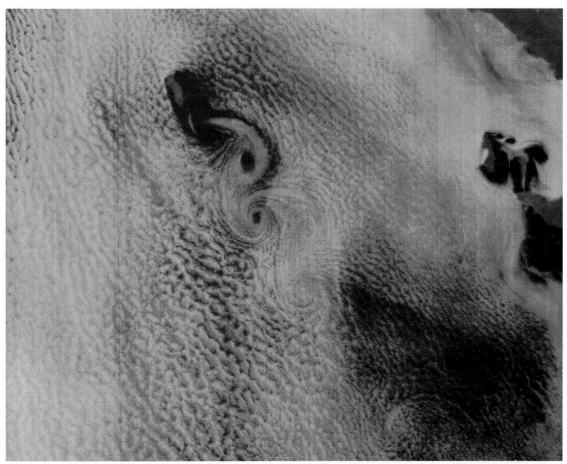

NSIDC, University of Colorado

LEE EDDIES—Lee eddies form in the clouds downwind from Guadaloupe Island, in the Pacific Ocean 300 miles south of San Diego. These eddies are similar to the Catalina Eddy that develops off southern California, but are much more photogenic because of the isolation of the island.

Its cyclonic winds are only about 10 miles an hour, but the rotational motion encourages the growth of thunderstorms. In spring and summer the Denver cylone becomes one of the state's favorite breeding grounds for tornadoes. An interesting situation will develop in the 1990s, when Denver's proposed new airport is built near the usual position of the cyclone.

Another weak cyclone sometimes spins off southern California, as northwest winds pass Point Conception and turn toward land. The center of the counterclockwise swirl is often near Santa Catalina Island, giving it the name "Catalina Eddy." Winds around this eddy don't blow any trees down, but they do tend to push the smog inland.

89

Richard A. Keen

LENTICULAR CLOUDS—The streamlined features of these lenticular clouds over Mount Evans, Colorado, are strongly suggestive of the strength of the winds (blowing right to left) that created the clouds.

Richard A. Keen

BANNER CLOUD—High winds produce a banner cloud in the lee of Mount Rainier, Washington.

The smaller a mountain, the smaller the eddies. Hills may kick off eddies just a few feet across. These little swirls are commonplace on windy days in the mountains, picking snow off the ground and spinning it around for a few seconds. Usually, these "snow devils" aren't very strong. However, the eddies get their energy from the wind that creates them, and if these winds are hurricane-force, the eddies may be even stronger. Chinook windstorms have spawned swirls powerful enough to lift sheds and small planes off the ground.

Eddies big and small doubtlessly form downwind from every mountain and hillock in the West. Only the lucky few that pass through a populated area or over a wind gauge ever get to tell their story. Keep your eyes and ears open as you watch the weather—these swirling winds are everywhere.

DUST DEVILS

They dance and twirl across dry valley floors, spinning dust around and around and frightening jackrabbits. Sometimes they look like tornadoes, but in some ways they're more like midget hurricanes. This is why dust devils are so fascinating to watch. Where else can you see a miniature storm system, compressed so small that you can run right through it?

A flat bit of ground and a sunny, calm day are the main ingredients to make a dust devil. The stage is set a few hours before noon, particularly

91

during the late spring or early summer, when the sun beats mercilessly from a cloudless sky onto the parched earth.

Air warms up during the day because of the heat it gets from the sunlit ground. So it's not surprising that the ground heats up earlier and eventually gets hotter than the air above it. It also helps to have dry ground, so you don't waste solar energy evaporating moisture. At noon, when solar heating is at its greatest, the ground may get as much as 80 degrees hotter than the air. In 1972, a ground temperature of 201 degrees was measured in Death Valley, while the air was a brisk 128 degrees. One should always wear shoes in Death Valley.

Right next to the ground—an inch or two above the surface—the air can get nearly as hot as the ground. Farther up, the air gets less hot, and so on. So you end up with a thin layer of extremely hot air near the ground and cooler air aloft. At some point, a bubble of this hot air rises up into the cooler layer. More hot surface air rushes in to take the place of that pioneering hot bubble, and an upwardly mobile stream of hot air develops. These currents can go up thousands of feet, where glider pilots (once again) know them as "thermals." Thermals are also the favorite haunts of hawks and other birds looking for a lift.

Some thermals go straight up, while others—for a variety of reasons—start to spin. Thermals are much too small to be twisted by the same Coriolis effect that gets larger cyclones spinning; more likely, rising currents pick up small eddies downwind of small bumps in the ground (little lee eddies). It has even been suggested that coyotes and rabbits start the small swirls that grow into dust devils. In any event, the spinning goes faster as more hot air gets sucked into the base of the thermal. Mix in some dust, leaves or any other debris, so you can see the swirling column of air, and there's your dust devil!

A typical western dust devil may be 20 feet across and 100 feet tall, but in extreme desert areas such as Death Valley and the Gila Desert near Phoenix, the dust column may rise half a mile high. Most dust devils last a few minutes at most, but the giant ones can live much longer. Some 2,000-footers have been followed for up to seven hours as they wandered 40 or so miles across Utah's Bonneville Salt Flats. These occasional giants are probably about as large as dust devils ever get—on Earth. The deserts of the West pale in comparison to the eternally rainless plains of Mars, where orbiting spacecraft have spotted monster dust whirls half a mile *wide* and three miles high.

The funnel of dust may look like a tornado, but its winds are rarely destructive. Tornado winds have been measured at over 200 miles per hour, while typical dust devils spin at one-tenth that speed. Winds have reached as high as 90 m.p.h. in some desert giants, but that's about tops for a dust devil. Upward wind speed is likewise smaller. While tornadoes have been known to lift cows and cars off the ground, dust devils prefer smaller game, such as tumbleweeds and kangaroo rats. Yes, kangaroo rats have flown in dust devils. To keep such critters airborne requires a vertical wind of 30 miles per hour.

At the same time dust and kangaroo rats may be swirling around the edge of the dust funnel, it's quite calm inside. In the calm center, air may actually be sinking slowly back to the ground—a feature strongly reminiscent of the "eye" of a hurricane! Another resemblance between the humble dust devil and the greatest storm on earth is the spiral of warm air into the center, where it rises and streams out the top of the whirlwind. Of course, there are differences. Dust devils are dry, while hurricanes thrive on moisture. There's also the obvious fact that you can fit billions of large dust devils inside a small hurricane.

Another class of dust devils can get a bit more serious than the little whirligigs that dot desert valleys on summer afternoons. These whirlwinds are often triggered by gust fronts from thundershowers. Acting much like a cold front, a gust front runs into hot air and gives it an upward shove. The resulting whirlwind—do we really want to call them dust devils?—can grow 1,000 or more feet high, and pack the wallop of a minor tornado. However, they lack the funnel-shaped cloud that tornadoes are famous for. More of a cross between tornadoes and dust devils than anything else, these hybrid creatures have been called "eddy tornadoes." Souped-up dust devils have been spotted by weather watchers across Colorado's High Plains and the Arizona desert, and no doubt many other spots in the West. Just remember to be careful with "dust devils" if there's a thunderstorm nearby!

Now we turn to some local winds, which, while they lack the drama of chinooks and dust devils, can still strongly affect the weather of the West.

SEA BREEZES

The most remarkable phenomenon of the weather there is the summer coast wind and its attendant mists. This seems to be due solely to the proximity of the districts of great heat and sudden rarefaction on the land, to the cold mass of waters off this coast, and its refrigerated surface atmosphere. A maximum day temperature of 110 degrees is often experienced at Fort Miller, a point in the San Joaquin Valley, when at the same time off Monterey and San Francisco the sea and sea wind are at 55 degrees. Such extreme contrasts existing at sea level and not far apart must be expected to originate violent winds, and it is only wonderful that they are not more severe at the passes giving access to the interior.

Lorin Blodget, in his 1857 *Climatology of the United States*, was well aware of the incredible contrast of temperature between California's coast and its Central Valley. Nowhere else in the United States, and possibly the world, do such extreme differences occur with such regularity. Blodget correctly supposed that this contrast would drive a wind inland from the sea—a "sea breeze."

Sea breezes blow Blodget's "refrigerated" coastal air inland to replace the hot air that rises over the interior—by now a familiar cycle to readers. This refrigerated air layer is rarely more than a few thousand feet deep and has a hard time getting past the 2,000- to 4,000-foot-high coastal ranges. There's only one place where the chilly air can pour relatively unimpeded into the Central Valley—the Golden Gate. This gives San Francisco its famous natural air conditioning in the summer.

The cooling effect fades rapidly as the oceanic air streams inland. On a typical July afternoon, downtown San Francisco averages only 64 degrees, but by the time the cool air reaches the city's airport—only seven miles inland—it has warmed to 72 degrees. Twenty-five miles inland, Walnut Creek averages 87 degrees, and when Tracy, 50 miles from shore at the edge of the Central Valley, gets the sea breeze it has warmed to a tepid 95 degrees. Deep into the valley, beyond the reach of the cooling breeze, Red Bluff and Fresno average a full 100 degrees on a July afternoon.

The persistent heat of the Central Valley virtually assures that sea breezes will cool the city by the bay from May through August, when most parts of the country are suffering their hottest weather. By September the heat of the Central Valley begins to moderate, and the sea breeze abates. Easterly winds from the land break through to the coast, and seaside towns even experience occasional heat waves. September and October are San Francisco's warmest months of the year!

Sea breezes on a smaller scale also ventilate the Los Angeles basin. Milder water leads to a weaker cooling effect, but it's still enough to chill a July afternoon on the pier at Santa Monica to 75 degrees while Canoga Park, 15 miles inland, averages 95 degrees. The wind from the sea also clears smog from the coastal points, but, unlike the

DUST DEVIL IN THE GILA DESERT NEAR PHOENIX, ARIZONA—*The land grader in the foreground gives some idea of scale. Clouds of dust are swirling around a tubular-looking core, where the winds are strongest. The air is nearly calm at the center of the tube.*

Sherwood B. Idso

David O. Blanchard

OREGON DUST DEVIL—A more delicate dust devil picks up dust from a roadside field in Oregon's Willamette Valley. The core of this whirlwind is only a few feet across, but the corkscrew pattern of upwardly swirling dust is clearly visible.

Santa Ana, wafts it back up the valley to places like San Bernardino, Riverside and Cucamonga. As a result, these communities at the eastern end of the Los Angeles basin frequently suffer the worst smog in the area.

The opposite of a sea breeze is a land breeze. At night, when the land may get colder than the ocean, the winds blow out to sea. One effect of this change is to bring more bugs to the beach at night—yet another reason the crowds leave when the sun goes down!

MOUNTAIN AND VALLEY WINDS

Those who live far from the ocean can take heart—they too have their version of the sea breeze. These winds are driven by the uneven heating and cooling of mountain slopes and valley floors, and so we call them "mountain and valley winds." Because these winds are as varied as the shapes of the terrain that cause them, let it suffice here to point out a few generic features.

Mountain and valley winds rely on the way air right next to the ground heats up faster by day and cools off faster at night than air farther off the ground. Remember this from dust devils? During the day, warm, light air near the ground drifts up hills and the sides of valleys, to be replaced by cooler air that sinks down the center of the valley. The warm air currents rising above the hills can lead to clouds and, sometimes, thunderstorms.

At night the opposite happens. Layers of cool air that form on hillsides slide into the valleys, creating air currents commonly called "drainage winds." If the hill is steep enough, drainage winds may start right after sunset, but on gently sloping terrain the cold air may have to build up for several hours before it starts moving. Being heavier, the cold air eventually finds itself in the lowest place around, which is the valley floor. This is why night temperatures are almost always colder in valleys than on mountain tops.

If you live in a valley or near a mountain, take note of how the wind direction changes from day to night. No matter where you live, the surest way to appreciate these fickle little winds is to camp in the mountains. When that first cool breeze slides down the valley during your evening meal, you'll know it.

INVERSIONS

The end result of nighttime valley breezes is often a valley full of cold air. If the breezes are doing their job, the coldest air is right at the bottom. This is an *inversion*, from a Latin word roughly meaning "upside down." Since air cools as it rises, in most situations the atmosphere is colder aloft than it is near the ground. With inversions, however, the atmosphere gets warmer with altitude.

This inverted atmosphere has one particularly nasty feature: if you give the cold heavy air near the ground an upward kick, it will come right back down. It's not easy to get cold air out of a valley. Whatever we put into the air—exhaust, smoke, dust and the like—stays right there with us. Virtually all episodes of heavy pollution come from the concentration of pollutants in a thin, inverted layer of air near the ground.

Inversions don't just happen in valleys at night. Los Angeles' famous smog is kept in by the shallow layer of cool air that drifts in from sea. In places like Denver and Salt Lake City the cold layer can be the remnants of an arctic cold wave. North or south, inversions remain until either the sun heats the cold layer into oblivion or a strong wind blows the mess away.

SOME BIG LOCAL WINDS

May 29, 1902—An intense dust devil, perhaps one of the tornado-dust devil hybrids, destroyed a livery stable in Phoenix. The next day another whirlwind unroofed a store.

April 21–23, 1931—Dry easterly winds downed trees on the western slopes of the Cascades, in a Pacific Northwest version of the Santa Ana. Chinook winds from the same storm system blew 11 freight cars off their tracks in Utah.

September 8, 1943—An inversion led to the first widespread smog in the Los Angeles basin.

October 30, 1959—A Wasatch chinook, with gusts exceeding 100 m.p.h., did millions of dollars' damage in Utah.

November 5–6, 1961—Extremely dry Santa Ana winds following a record hot summer fanned brush and forest fires across the Bel Air section of Los Angeles.

January 25, 1962—A Montana Air National Guard C-47 crashed near Wolf Creek, killing six, including the state's governor. Severe mountain wave turbulence had torn one wing off the plane.

April 3, 1964—Easterly chinook winds along the Wasatch Mountains blew trucks over between Salt Lake City and Ogden, Utah, and turbulent winds aloft flipped over a crop duster. Fortunately, the pilot regained control of his plane.

March 12, 1968—A Washoe Zephyr swept the eastern slopes of the Sierra Nevada, damaging buildings and trees in Reno, Nevada.

January 7, 1969—Chinook winds gusting to 130 m.p.h. in Boulder, Colorado, sent roofs and parked airplanes into flight, and damaged half of the city's houses.

September 22–29, 1971—Hot, dry Santa Ana winds led to widespread fires from southern California to the Oakland–Berkeley area. More than 500 homes and half a million acres were consumed.

March 6, 1972—Powerful chinook winds, gusting to 90 or 100 m.p.h., blasted Livingston, Montana, and Lander and Cheyenne, Wyoming, with $1 million in damages to roofs, airplanes and other typical targets.

November 12, 1973—Strong winds, intensified and localized by mountain waves, toppled 10,000 trees in a narrow, 15-mile-long swath east of the Teton range in Wyoming.

July 4, 1978—Out of a clear blue sky, a potent dust devil tore the roof off a mobile home in Salida, Colorado, sending rafters flying 200 feet into the air.

February 13, 1979—Southwesterly winds from an offshore storm spawned a lee eddy east of the Olympic Mountains. The localized 100 m.p.h. winds sank the Hood Canal pontoon bridge.

December 4, 1979—From Montana to Colorado, chinook winds as high as 119 m.p.h. rolled over airplanes and mobile homes. Twenty-three freight cars were blown off the tracks near Cheyenne.

January 17, 1982—Powerful chinook winds, measured as high as 140 m.p.h., raked the eastern foothills of the Colorado Rockies. In Boulder, utility poles were snapped in half, and damage totaled $17 million. The high winds raised Boulder's temperature from two degrees to 50 degrees in one hour. A week later, another windstorm generated tornado-force eddies, one of which popped the windows out of 100 cars parked at a shopping center.

October 9, 1982—Sixty-mile-per-hour Santa Ana winds fanned flames that destroyed nearly 100 homes near Los Angeles. One fire burned a 15-mile path across the Santa Monica Mountains to the sea.

April 4–5, 1983—Two days of chinook winds raked the western slopes of Utah's Wasatch Mountains, blowing down high-tension power lines, overturning mobile homes and rolling trucks off Interstate 15. Damage totaled $8 million.

December 24, 1983—Cold "bora" winds as strong as 100 m.p.h. descended the western slopes of the Cascades, flattening several homes near Enumclaw, Washington, while easterly gales funneling through the Columbia Gorge downed trees and power lines. Total damage reached $27 million.

HOW HOT, HOW COLD

The West has long been renowned for its diverse weather. In *Roughing It*, Mark Twain recounts with pithy eloquence his impression of the first desert he had ever seen, the Great Salt Lake Desert:

The sun beats down with dead, blistering relentless malignity; the perspiration is welling from every pore in man and beast, but scarcely a sign of it finds its way to the surface—it is absorbed before it gets there; there is not the faintest breath of air stirring; there is not a merciful shred of cloud in all the brilliant firmament; there is not a living creature visible in any direction whither one searches the blank level that stretches its monotonous miles on every hand; there is not a sound—not a sign—not a whisper—not a buzz; or a whir of wings, or distant pipe of bird—not even a sob from the lost souls that doubtless people that dead air.

Mark Twain found the climate of Lake Tahoe, his destination, to be a world apart from that of the salt flats:

Three months of camp life on Lake Tahoe would restore an Egyptian mummy to his pristine vigor, and give him an appetite like an alligator. I do not mean the oldest and driest mummies, of course, but the fresher ones.

Twain's comment that "if you don't like the weather, wait a minute," was said about the climate of New England, but others have noted the rapid changes that can occur in the West. During the course of his transcontinental survey for the Central Pacific Railroad in 1853, Captain Beckwith remarked:

Several times during the day (Sept. 5) we experienced very sensibly the sudden changes of temperature to which high altitudes in mountain regions are subject from a passing storm or a change of wind—our thick coats being at one moment necessary to our comfort and the next oppressive.[3]

These narratives were written before there was much in the way of actual weather records in the West. Lewis and Clark and Zebulon Pike made weather observations during their explorations of the West in 1804–07. The first western weather station may have been managed in 1821 by the British at Fort George, near the present town of Castle Rock, Washington. Early weather stations were spotty and short-lived. Weather data from a network of consistently reporting stations didn't begin until after 1846, when most of the West became United States territory and the army began taking observations at its newly emplaced forts and posts. It didn't take long for these record keepers to realize how extreme their climate could be.

During the summer of 1853, a high temperature of 121 degrees was recorded at Fort Miller, in California's San Joaquin Valley. At the time, this was probably the hottest temperature ever recorded in the world—the Old World record was 117 degrees, at Esna, Egypt. Up north in Montana in 1885, the alcohol (mercury freezes at

WESTERN WEATHER EXTREMES

Wynoochee Oxbow—ave. 144" rain per year

Tillamook Forest— max gust 170 m.p.h. (October 12, 1962)

Mt. Rainier—1122" snow (1971-72)

Browning fell 100° overnight

Glasgow—all-time temperature range, 113 to -59°

Rogers Pass— -70° (1954)

Spearfish—rose 49° in 2 minutes

Eureka—never warmer than 86°

Cheyenne—averages 10 hail storms per year

driest state—ave. precipitation 8.68"

Peter Sink— -70° (1985)

Silver Lake—76" of snow in one day

Death Valley—134° 1.68" rain a year

Cimarron—110 thunderstorm days per year

San Diego—has never had measurable snow

Yuma averages 91% annual sunshine

40 below, so alcohol thermometers are used) plunged to a frigid 63 below zero at Poplar River. The army beat that three years later with a 65-below reading at Fort Keogh (near Miles City). Before these Montana low temperatures, the record cold readings for the United States had been in Vermont, New Hampshire and Maine. From 1885 on, the all-time temperature extremes for the United States (except Alaska, of course) belonged to the West. All these extremes have been exceeded in the present century. We'll get to these later.

The weather of the West holds more surprises than simple highs and lows of temperature. Some changes of temperature are equally remarkable—not just changes that occur over time, but also differences between places just a few miles apart (vertically and horizontally). There are locations where the degree of moderation in climate is as notable as the extremes are elsewhere. We'll go through these climate oddities one at a time and give some explanation as to why they occur. You'll find that the causes can be as interesting as the records themselves, because there's a lot to be learned from looking at these freaks of weather.

HOW HOT?

Since the 121-degree reading at Fort Miller, California, several other states in the Union have reported readings as high or higher. Nevada has recorded 122 degrees, and Arizona a blistering 127. During the Dust Bowl summer of 1936, highs of 121 degrees seared Kansas and—believe it or not—North Dakota. The honors go to Death Valley, California, which recorded 134 degrees Fahrenheit on July 10, 1913.[4] As a new world record, it didn't last long. Nine years later, a 136-degree reading was taken at the little oasis of Azizia, Libya. But Death Valley's 134-degree-day remains the nation's hottest, and its annual average of 77 degrees is the highest in the country.

On the other end of the heat scale, a 14-year-long record of temperatures (1874–88) atop Pikes Peak, Colorado, showed a maximum of only 64 degrees. This is probably the lowest extreme maximum of any western weather station with a record lasting more than a few years. Incidentally, the average yearly temperature on that famous mountain is 20 degrees, the lowest in the 48 states. Doubtless there are places—such as the top of Mount Rainier, Washington—where the maximum would be even lower had somebody been fool enough to stay up there keeping records for several years. (Temperature extremes for Death Valley, Pikes Peak and other locations in the West, along with other climatic data, are listed in Appendix 1.)

The reason for the difference between Death Valley and Pikes Peak is obvious. The Pikes Peak weather station stood at 14,134 feet above sea level, while the Death Valley station sits at 178 feet *below* sea level. Air taken from Death Valley and lifted 14,312 feet to Pikes Peak cools by 79 degrees due to the pressure drop alone. The observed difference of 70 degrees is pretty much what one should expect based on the altitude difference. This means that equally hot air can reach into most parts of the West at some time during the summer, and that it is primarily the altitude of the weather station that determines the extreme highest temperatures.

But what of Eureka, California? This coastal city in northern California has never been hotter than 86 degrees, and yet it is only a few feet higher than Death Valley. Much hotter readings have been observed at Fort Yukon, Alaska, above the Arctic Circle, and I have personally recorded the same 86 degrees at 9,000 feet in Colorado. These latter two places are far from any ocean; Eureka, on the other hand, sits right next to the Pacific and owes its moderation to the chilly offshore currents.

An interesting feature of the West Coast is that the coldest offshore water is around northern California near Eureka. Water here averages about

55 degrees, colder than to the south or north along the coast. The cold water results from the southbound California Current. Remember the Coriolis effect? It works on ocean currents as well as on winds, and tries to make the California Current turn to its right—i.e., offshore. To counter this effect, cold water from the deeper ocean surfaces along the shore, extending out about 25 to 50 miles. This phenomenon, called "upwelling" by oceanographers, brings a continuous supply of chilly deep water to the surface and maintains the low temperatures near Eureka. The cold water, in turn, chills the lowest 1,000 or 2,000 feet of atmosphere. Eureka sits in the cold layer all summer long, and its temperature rarely rises above 70 degrees.

HOW COLD?

The last we looked, Fort Keogh, Montana, had seen a low of 65 degrees below zero. Records, like rules, are meant to be broken. In 1933, the national low left Montana and moved into Wyoming (but just barely), when the thermómeter at Yellowstone National Park's Riverside Ranger Station dipped to 66 below. That record stood until January 20, 1954, when 5,470-foot Rogers Pass, Montana, 40 miles northwest of Helena, saw the temperature dip to 69.7 degrees below zero. Let's call it 70 below; it is still the national record.

In world competition, the West doesn't do as well in the cold category as it does in the hot one. Rogers Pass is still 57 degrees shy of the world record of 127 below, recorded at the Soviet Union's Vostok station, 11,000 feet up on the Antarctic Plateau. North America's low is 81 below on the runway at Snag, in the Yukon, and Alaska has seen 80 below.

On the high end of the low-temperature scale, Los Angeles has never chilled below 28 degrees. This means that every city in the West has experienced subfreezing weather. Even on the pier at Avalon, on Santa Catalina Island 26 miles

offshore from Los Angeles, the mercury has dipped to the freezing mark. The only city in the United States (outside of Hawaii) that has never frozen is Key West, Florida.

It takes more than elevation to explain the difference between Rogers Pass and Los Angeles. The 5,400-foot separation in altitude accounts for only 29 degrees of the 98-degree observed difference in minimum temperature. Unlike the situation with extremely hot air in the summer, which at times may spread fairly uniformly across the West, there is an important difference in the cold air masses that reach the two locations. The air that invades Montana in winter is much colder than that which moves into southern California.

Cold air masses responsible for the low temperatures at any location in the West form over the arctic regions of northwest Canada, Alaska and Siberia. While these areas receive virtually no sunlight during winter, their snow-covered surfaces radiate heat off into space. As a result, any air that passes over the Arctic finds itself losing heat and getting colder. The longer it stays, the colder it gets. Eventually these refrigerated air masses are taken south by cyclones forming along their southern boundaries.

The normal west-to-east flow of jet streams across the western states brings a stream of relatively mild air from the Pacific Ocean, and keeps arctic air from moving south. As the arctic air continues to chill, however, the temperature difference grows between the Arctic and the tropics. With this temperature difference, the strength of the jet must also increase. When the jet blows too fast, it starts to buckle. The buckling takes the form of troughs and ridges in the upper atmosphere and of fronts, cyclones and anticyclones in the lower atmosphere. And the frigid mass of arctic air begins to move south.

The path of the cold air outbreak depends on where the buckles form on the jet. A common pattern is for the ridge to develop over the Yukon and the trough over the Great Lakes. This pattern brings the cold wave into the Midwest and East. Less frequently, an Alaskan ridge and Rocky Mountain trough deliver the arctic air into the Rockies. If the trough forms over the West Coast, the cold wave strikes the Pacific Northwest and perhaps moves into California. This last pattern is the rarest of all, which explains why winters are so mild along the coast.

Once a cold wave has overrun the West, local terrain may play some tricks with the temperature. At the Pikes Peak weather station, the lowest recorded temperature was 39 below. Yet 85 miles away and 5,000 feet lower, Taylor Park has been down to 60 below zero. Three thousand feet lower—nearly two miles below Pikes Peak—the town of Maybell, Colorado, has recorded the state's low of 61 below. Does it really get colder at lower elevations?

The answer is yes and no. Cold air is heavy, and extremely cold air tends to hug the ground. The coldest air masses moving south from the Arctic often extend upward only two or three miles. Within the depth of the air mass, air near the top is colder than that near the surface, due to the familiar drop-off of pressure. Once the cold air settles in over the mountains and valleys of the West, however, locally chilled air has a habit of sinking into the valleys at night. If the valley has a thick blanket of snow on the ground (which effectively insulates the air from the residual heat in the ground, and vice versa), a thin layer of air near the ground can cool off extremely rapidly after sunset, leading to very low overnight temperatures. The day of its 70-below reading, Rogers Pass had five feet of snow on the ground. The next morning, as the sun heated the shallow surface layer of cold air, temperatures rose as fast as they had fallen the night before. Rogers Pass warmed back up to near zero the afternoon following its record low reading.

A shallow cold air mass may sweep into the lowlands, but leave the mountaintops in mild Pacific air. A deeper air mass engulfs mountains and valleys alike. During the day, the mountains, being higher up, are colder than the valleys. At night, air locally cooled by radiation sends the valley bottoms into the icebox. A high valley gets colder air to start with, leading to even colder readings at night. The ideal conditions for extreme cold temperatures are clear, calm nights in high-altitude, mountain-rimmed valleys with deep snow on the ground.

UPS AND DOWNS

Temperatures in the West have ranged from 134 above to 70 below, an incredible 204-degree spread. But Death Valley is a long way from Rogers Pass. What are the greatest (and smallest) temperature ranges ever recorded at one place? What are the quickest changes ever recorded?

Most of these extreme temperature-change records come from northeastern Montana. Montana's fame stems from its location relative to the average positions of the fronts and air masses that affect the West. Northeastern Montana is closer to the main track of the arctic air masses that invade the United States from the north, and thus is subject to the coldest of these air masses. In winter, western Montana usually remains in the mild air blowing in from the Pacific Ocean. Northeastern Montana is therefore the frequent battleground of the two contrasting air masses, and the battles can lead to some spectacular temperature changes. In summer, Montana is subject to the same extremely hot air mass that covers the rest of the West. While Montana's extreme highs may equal those recorded anywhere else, its extreme lows can be much lower. Record ups and downs include:

Greatest all-time temperature range—172 degrees, Glasgow, Montana, from 59 below (February 1936) to 113 above (July 1900).

Greatest range in one year—171 degrees,

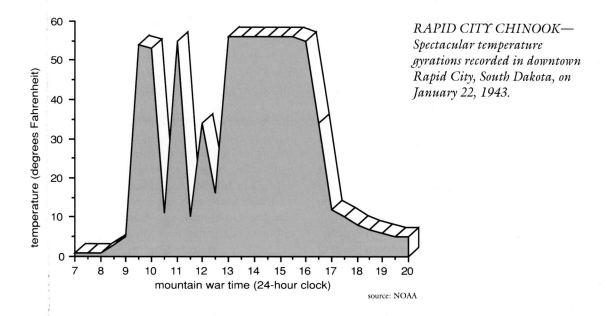

RAPID CITY CHINOOK—
Spectacular temperature
gyrations recorded in downtown
Rapid City, South Dakota, on
January 22, 1943.

source: NOAA

Glasgow, Montana, in 1936, from 59 below (February 15) to 112 above (July 18).

Greatest range in one day—100 degrees, Browning, Montana, January 23–24, 1916. Temperature fell from 44 above to 56 below in less than 24 hours with the passage of a cold front.

Greatest warming in two minutes—49 degrees, from 4 below to 45 above, at Spearfish, in the Black Hills of South Dakota, from 7:32 to 7:34 a.m. on January 22, 1943.

The most remarkable temperature fluctuations occur east of the Rockies when the thermometer happens to be situated near the top of a shallow cold air mass. The dense, cold air may act like water on a beach. As it washes in and out, the temperature jumps up and down nearly as fast as the thermometer can record it. At Great Falls, Montana, on January 11, 1980, the mercury rose from 32 below to 15 above—a 47-degree change—in seven minutes. Havre, Montana, once reported a warming of 26 degrees in 45 seconds.

The most celebrated case of wild tempera-

ture fluctuations began with the record two-minute warming at Spearfish. An hour and a half later, the temperature had slowly risen to 54 degrees. Suddenly, the arctic air sloshed back in, dropping the temperature 58 degrees—from 54 above to 4 below—in 27 minutes. Imagine trying to dress for a day like that!

Less dramatic are the records for the *smallest* changes of temperature. There have been many instances of 24-hour temperature fluctuations of zero degrees—no change all day. But for the smallest range of all-time extremes at any one location, we return to that moderate city by the sea, Eureka, California. In 100 years of record, the temperature has never exceeded 86 degrees or fallen below 20 degrees. The 66-degree spread is the smallest in the West. In the 48 states, only Key West, Florida, with a range of 56 degrees, has a more uniform climate. Over the course of a century, Eureka has experienced a change of temperature scarcely greater than what Spearfish has seen in half an hour.

103

SKYWATCH

FROM HERE TO THERE

Some of the temperature differences between places a few miles apart can be as impressive as the changes over short periods of time. Differences exceeding 40 degrees are common between Pikes Peak and Colorado Springs, 12 miles east and 8,100 feet lower. If there were a weather station on Mount Whitney, California, we might see temperature differences of 70 or 80 degrees over the 80 miles and 14,600 vertical feet that separate it from Death Valley. However, this is cheating. We expect it to be much colder on mountains.

The largest *horizontal* temperature differences occur along fronts. On the eastern plains of Montana, Wyoming and Colorado, temperature changes of 40 to 50 degrees over as many miles happen several times each winter. The wildly gyrating temperatures at Spearfish were caused by slight shifts in the location of an incredibly sharp front. At times, this front cut towns and even city blocks in half. In Rapid City, South Dakota, weatherman Roland Hamann reported that "at 11 a.m., on the east side of the Alex Johnson Hotel, winter was in all its glory, while around the corner on the south side, not 50 feet away, spring held sway, only to be swept away in a flash by the sting of winter, and then to return." Motorists driving across town reported fogging, frosting and even cracking of windshields as they drove between the air masses. In one gully, the temperature on the bank was measured as 40 degrees warmer than at the creek bed, just 20 feet below.

At one time, the Black Hills town of Lead reported 52 degrees, while Deadwood, *two miles away,* froze at 16 below! Fortunately, this remarkable front spent the day in an area populated by weather enthusiasts, so it was well-documented. Generally, however, weather stations in the West are spaced too far apart to catch such incredible differences of temperature.

Big temperature differences usually don't last very long because cold fronts and warm fronts move. One front that sits around for months on end is the separation between the cold, foggy air along the California coast and the hot, dry air inland over the San Joaquin and Sacramento valleys. This front is a near-permanent feature during the summer, and the temperature changes across it can be every bit as large as those in a wintertime cold front in Montana. When Fort Miller, in the San Joaquin Valley, recorded its world-record 121 degrees in July of 1853, Monterey, on the coast 130 miles to the west, never got above 70 degrees. For the entire month, Monterey averaged 45 degrees cooler than Fort Miller.

Over a century later, in August 1981, the town of Red Bluff, at the northern end of the Sacramento Valley, recorded its all-time high temperature—also 121 degrees. At the same time, the now-familiar city of Eureka shivered in the fog and in the heat of the day only made it to 61 degrees. That's a 60-degree difference only 100 miles apart. The boundary between these vastly differing air masses moves very little and from June to September can usually be found somewhere in the Coast Ranges of California. With its 50-degree temperature contrasts, it may very well be the most persistent front in the world.

It may be academic, perhaps, but the greatest temperature difference ever recorded across the entire West—that 11-state region that is the subject of this book—probably occurred at 2 a.m. on January 20, 1954. The weather observer at Rogers Pass read 70 below from his thermometer. At that same moment in the balmy desert Southwest, Yuma, Arizona, read 55 degrees. Over a distance of 1,000 miles, the temperature difference was 125 degrees.

What is the *least* temperature difference the West has ever seen? Some help in figuring this one comes from the fact that the farthest corners of the region, San Diego and Havre, Montana, have exactly the same average temperature—70 degrees—during July. Consequently, we may expect that during the summer, the difference between these two points 1,250 miles apart may often amount to precisely nothing!

In the region called the West there are places where the temperature has fallen 100 degrees in a day and places where it hasn't nudged a degree. The West harbors the hottest and coldest places in the 48 states, as well as the most and least variable climates. At any moment the weather may range from arctic to tropical across its 11 states, or it may be identical at locales 1,000 miles apart. This, in a nutshell, is the varied weather of the West.

Following is a list of some of the extremes of heat and cold to visit the West over the past century or so. The selection was a difficult one, since nearly every year someplace in the West gets a memorably hot or cold spell. The events chosen here are based as much on their extent as on their severity.

GREAT WESTERN HEAT WAVES

Summer 1913—In July, Death Valley baked at 134 degrees, the hottest temperature ever recorded in the New World. Two months later, coastal San Diego reached 110 degrees.

July 1934—Idaho and New Mexico recorded their hottest days ever, at 118 and 116 degrees respectively.

Summer 1937—Yuma, Arizona, broiled at or above the century mark for 101 days in a row. In July, the heat reached into Montana, hitting 117 degrees at Medicine Lake, just 35 miles from the Canadian border. Across the border, the mercury soared to 113 degrees at Yellow Grass, Saskatchewan, an all-time high for Canada. This was in the middle of the Dust Bowl years, 1933–39, the worst drought of record in the Great Plains. The heat often extended into the West, and many places had their hottest days, months or summers of record during this period. Amazingly, many of these same places recorded their coldest winters during the same decade. It was a period of

extreme variability of western weather.

June–July 1954—One of the most extensive hot spells in history scorched the nation, coast to coast. Six states saw their hottest days ever, including 122 degrees at Overton, Nevada. In the West, records were broken as temperatures soared to 99 at Ely, Nevada, 100 at Cheyenne, Wyoming, 104 at Boulder, Colorado, and Casper, Wyoming, and 106 at Sheridan, Wyoming

June–September 1955—Torrid times continued, with another coast-to-coast heat wave. Seattle saw 100 degrees, an all-time high for the city. Los Angelenos withered under an eight-day spell of 100-degree-plus weather, with the mercury reaching 110 degrees on September 1. For the next six years, unusually hot summers were the rule for southern California.

Summer 1959—Desert heat at its finest. Death Valley lived up to its reputation, with 126 days in a row at 100 degrees or hotter, topping out at 125 degrees.

July–August 1961—A searing summer in the North. Lewistown, Montana, soared to 115 degrees and, at inappropriately named Ice Harbor Dam, Washington, a 118-degree reading was a record for the state. Even on the coast, Astoria, Oregon, topped out at 100 degrees, and 106 was reached at San Francisco's airport. Widespread forest fires broke out all across the West. As the heat continued into November, fires struck the Los Angeles area.

August 1969—Another hot one in Montana. Helena's hottest day (105), and a toasty 110 at Havre. Havre's high is all the more impressive when you consider that residents shivered at 52 below a half year earlier.

August 1981—The red-hot conclusion to a searing summer. All-time high temperatures were recorded at Red Bluff, California (121), Portland (107), Olympia, Washington (104), and Quillayute, on the Pacific shores of Washington's Olympic rain forest (99). The persistent heat made

R. C. Rothermel, U.S. Forest Service

FOREST FIRE—The heat of the Sand Point Fire, just east of Great Falls, Montana, in July, 1985, created this mushroom-shaped cumulus cloud.

it Phoenix's hottest summer ever.

September 1983—The coastal cold front pushed out to sea, and Eureka, California, toasted briefly at 86 degrees—an all-time record.

June–September 1988—A summer reminiscent of the Dust Bowl days scorched the country from coast to coast. Two cities recorded their hottest days ever—Cleveland (104 degrees) and San Francisco (103), while Death Valley peaked at 127 (not a record). In August dozens of lightning-ignited forest and range fires broke out across the West, with the largest fires consuming much of Yellowstone National Park. Smoke from the Yellowstone fires rose to 30,000 feet and dimmed the sun hundreds of miles away.

June 1990—Temperatures more appropriate for a cookbook than a weather report seared the southwest. Phoenix topped out at 122, 4 degrees higher than the previous record, and Tucson reached 117 degrees. The hot weather at Phoenix forced the closing of the international airport—airplanes couldn't take off in the heat-thinned air!

GREAT WESTERN COLD WAVES

January 1854—Cold engulfed the entire West, sending the temperature to 1 below at Steilacoom (near Tacoma), Washington, and 25 at San Francisco.

Winter 1886–87—More than half of Montana's cattle population froze to death during a bitter winter that brought several blizzards and temperatures reaching 60 below. The disaster put an end to open-range cattle raising in the state.

January 1888—One of the Pacific Northwest's greatest cold waves, with 2 below in Portland, Oregon, and 65 below in Montana, and all-time record lows in Spokane, Washington (30 below), and Eureka, California (20). Portlanders could walk across the frozen Willamette River for 10 straight days.

January 1893—Another Pacific Northwest cold snap, with downtown Seattle's all-time low of 3 above, and 7 degrees offshore at Tatoosh Island, Washington.

February 1899—The most severe and widespread cold wave ever to freeze the United States. Subzero readings were reported from every state, ranging from 61 below at Fort Logan, Montana, to 2 below at Tallahassee, Florida. All-time records were set at such diverse locations as Billings (-49) and Miles City (-49), Montana; Greeley, Colorado (-45); Brownsville, Texas (+12) and Washington, D.C. (-15).

January 1913—Freezing weather swept into the normally mild Southwest, setting all-time lows at San Diego (25), and Phoenix (16) and Tucson (6) Arizona. Fifty below at Strawberry Tunnel, Utah, remains the coldest officially measured in that state.

January 1930—Temperatures fell to 52 below in Oregon and Montana and to 57 below in Wyoming. The Columbia River was frozen for two weeks. It was Wyoming's coldest January ever, but was followed by the *warmest* February on record.

February 1933—Extreme cold over the northern Great Basin and Rockies set all-time low temperatures for Wyoming (-66), Oregon (-54), and Salt Lake City (-30). The Wyoming low held the national record for 21 years.

February 1936—A series of cold waves over the plains east of the Rockies set all-time record minimum temperatures at Denver (-30); Lander, Wyoming (-40); and Great Falls (-49) and Glasgow (-59), Montana.

January 1937—The Southwest shivered as all-time cold records fell from Elko, Nevada (-43), to Yuma, Arizona (+22). Cold records for the state were set in Boca, California (-45), and San Jacinto, Nevada (-50).

January 1943—Sixty below set a state record for Idaho. The return of mild Pacific air following the cold outbreak led to the spectacular gyrations of temperature at Spearfish, South Dakota.

January 1949—In terms of average temperature, it was the coldest winter most of the West has ever seen, and right in the middle of it, a sharp cold snap set record low temperatures from Pocatello, Idaho (-30), to downtown Los Angeles (28). Even on the pier at Santa Catalina Island, off Los Angeles, the temperature fell to 32 degrees.

January 1954—Seventy below zero at Rogers Pass set a new national record, but the cold was short-lived. At nearby Helena, where the low was 36 below, temperatures soared to 54 above—a recovery of 90 degrees—in just two days. Statewide, the winter as a whole averaged 4 degrees warmer than normal.

January 1963—The first of a series of arctic blasts to strike the Rockies this month was the worst. West Yellowstone saw 60 below, and Denver recorded an afternoon maximum of 15 below. Las Vegas froze at an unprecedented 8 degrees.

December 1968–January 1969—A late December cold wave struck the Northwest with 50 below in Moscow, Idaho, and 11 above on the coast at Astoria, Oregon. In northern Montana, the entire winter—December through February—averaged only 1 degree above zero, the coldest winter ever observed anywhere in the West. At Havre, the mercury remained continuously below zero for the last 17 days of January, and Hinsdale dropped to 55 below.

December 1983—Christmas was chilly this year—52 below at Butte, Montana, where the entire month averaged 6 below zero. Denver's temperature remained below zero for nearly five days, while parts of Montana suffered twice as long. Meanwhile, the Southwest remained unusually *warm* all month.

February 1985—Polar air and light winds sent thermometer readings way below zero over the intermountain region. Maybell, Colorado, saw 61 below, a new record for the state. It was a classic case of rapid cooling of a shallow layer of air near the ground—Maybell's temperature recovered to a balmy 15 *above* by the same afternoon. Meanwhile, a research station in Utah's Wasatch Mountains recorded 69.9 degrees below zero. Had that reading been taken at a permanent weather station, it would be a new official record for the 48 states, breaking the Rogers Pass low by one-fifth of a degree!

February 1989—Alaska's second-coldest cold wave in history (-76 at Tanana) poured south into Montana, where Wisdom dipped to 52 below, and Colorado, where my backyard thermometer in Coal Creek read -36. Gale-force easterlies blew 7-degree air into Seattle, and after crossing the mild waters of Puget Sound, the cold air left 3-foot "sound-effect" snowfalls on the Olympic Peninsula. For much of the West it was the coldest February weather ever experienced.

EL NIÑO

Californians are becoming concerned that the dreaded 'el Niño' is again about to strike their shores.

This comment, from a 1986 television documentary, illustrates some of the apprehensions—and misconceptions—that westerners have about the weather upheaval known as "el Niño." Recent history makes it clear that the Niño phenomenon is of great concern and interest to western weather watchers. The winter of 1982–83 brought a Niño that has been rated the biggest in a century. California was slammed by a relentless series of Pacific storms, with winds and pouring rains washing away the shoreline and sliding mounds of mud onto highways. Many will never forget the television images of houses and hot tubs floating away in the surf. Heavy snows choked the Sierras and Rockies, and even the cacti of northern Mexico were dusted several times. So what is this "dreaded el Niño" that strikes the West?

Sometimes it is easier to tell what some-thing isn't rather than what it is, and el Niño is *not* a storm that strikes the West Coast. As a matter of fact, the phenomenon that is correctly called "el Niño" happens far from western shores, and has little if any impact on western weather. However, it's a catchy name, and in the absence of a better one, it appears that whatever it is that strikes the West will be called el Niño for some time to come.

In its original sense, *el Niño* is the name given by fishermen of coastal Peru and Ecuador to a warm ocean current that shows up every so often. Normally, the western coast of South America is swept by a northward current of cold water from the Antarctic. Along the immediate coast, the Southern Hemisphere Coriolis effect (which is the reverse of the northern effect) causes upwelling immediately offshore, keeping the water good and chilly. This is the Southern Hemisphere counter-part of the current that refrigerates the California coast, especially around Eureka, all summer long. The fresh supply of deep water brought to the

109

surface by this upwelling is loaded with organic nutrients, which feed the plankton and, in turn, the swarms of anchovies that populate the area. As a result, these Peruvian and Ecuadoran coastal areas are among the richest fishing grounds in the world, although in recent years they have been severely depleted by overfishing.

The coastal current and its attendant upwelling vary according to seasons. In December at the height of the Southern Hemisphere summer, the current usually weakens and the coastal water warms up a little. Some years the current ceases or reverses direction, and the water warms up a lot—as much as 10 degrees or more. In these years the nutrient-loaded upwelling ceases, dwindling the fish population. Since these warmings often begin around Christmas, the locals have dubbed the warm current el Niño. The Spanish translates to "the Child," referring to—without implying blame—the Christ child. Recent Niños occurred in 1957, 1972, 1976 and 1982. Combined with the overfishing that has taken place, these Niños have become increasingly disastrous for the local fishing industry, not to mention for the fish themselves and the birds that feed on them.

The phenomenon doesn't end with dead anchovies littering the beaches of Peru and Ecuador. The warm water may spread, reaching westward across the tropical Pacific and northward along the west coast of Mexico toward California and Oregon. At this point the warming becomes a global concern, since its worldwide climate effects can be enormous. However, out of due respect for the fishermen who coined the name, remember that the term "el Niño" really applies to the warming off the South American coast. It is, to say the least, stretching it a bit to apply the name to storms in California and the Rockies, but again, for lack of a better name, let's use it.

The variety and scale of the worldwide weather freaks that occurred during the 1982–83 Niño are truly impressive. Elsewhere during that Niño year, hurricane-force cyclones pounded the Gulf of Mexico and Atlantic coasts; however, in terms of temperature, the winter was exceptionally mild nationwide. Around the globe, torrential rains flooded desert areas of Peru and Ecuador, while droughts struck Brazil, Africa and Indonesia, and the life-giving rains of the Indian monsoon arrived late. A long, hot summer led to terrible brush fires that consumed thousands of square miles of Australian bush, along with several small towns. Many meteorologists—but not all, mind you—believe that all these weather disasters were related, and that the massive warming of the Pacific Ocean provided the common link.

Let's slow down a bit. Could all this bad weather really be the fault of some warm water? The last Niño before 1982 was in 1976, and the following winter was nothing at all like the storms of 1982–83. The nation's bicentennial was the year of a great western drought, especially in California. Advertisements exhorted people to "shower with a friend" to conserve water. Ski areas from California to Colorado suffered snow shortages. Meanwhile, in the central and eastern states it was one of the bitterest winters in history. It was the year of the savage "Buffalo blizzard" in upstate New York, and snow even fell on Miami. Nationwide, it was the driest winter in recent history, while 1982–83 was one of the wettest. What happened to all the storms striking the coast? Why was it so cold? Was the 1976 Niño different from that of 1982, or is it all a bunch of bunk?

The reason for the differences between 1976–77 and 1982–83 lies in the misuse of the term "el Niño." True, both years saw Niños off the coast of Peru. However, this relatively local phenomenon is not the part of the global upheaval that strikes the West. Rather, it is the warming of large areas of the rest of the Pacific Ocean—and especially the tropical Pacific—that upsets western weather. These warmings were quite different between the two years.

CHANGING THE WIND

To paraphrase certain stockbrokers, when the Pacific warms up, the world listens. Indeed, a warmer-than-usual tropical Pacific Ocean can disturb the most fundamental forces that drive the world's weather. The most powerful and persistent winds of the world—the subtropical jets of the Northern and Southern hemispheres—result from the upward and outward flow of air in the huge thunderstorm masses that cover Africa, South America and Indonesia. North of the equator, air streaming north from the upper levels of these thunderstorms turns into an eastward-blowing jet stream, thanks to the slower speed of the underlying ground at higher latitudes. The predominant patterns of the subtropical jet result from the locations of the thunderstorm masses, with the jet being farther north and stronger due north of the storms. South American thunderstorms strengthen the jet over eastern North America and out into the North Atlantic. This Atlantic jet blows away from the western states, so it is the next jet to the west—the one north of Indonesia—that is the most important for western weather.

Normally, when the tropical thunderstorms are dousing Indonesia, New Guinea and other islands at the western limits of the Pacific, the subtropical jet is strongest from eastern China to the western Pacific south of Japan. This jet stream is so persistent across most of the Pacific that during World War II, the Japanese used it to float balloon-borne bombs to the United States. However, the jet has a tendency to peter out by the time it reaches the West Coast, and often splits into two weak jets that skirt the northern and southern borders of the West. This kept most of the Japanese balloons from ever reaching the United States, although several did make it, causing several deaths and minor damage.

The thunderstorms that drive the subtropical jet are fed by latent heat evaporated from the surfaces of the warm tropical seas. These thunderstorms prefer to build up over the equatorial land masses, where the overhead sun is most effective at directly heating the ground and causing the rising currents of air that trigger the storms. Of the three major land masses along the equator, Indonesia is the smallest, being in reality a bunch of islands. Most of its surface area is water. Because Indonesia is surrounded by the most extensive and warmest of all the tropical ocean areas, it is the thunderstorm capital of the world.

When the tropical Pacific Ocean gets even warmer than usual, as happens during some Niños, a major shift in the position of the Indonesian thunderstorm mass can take place. Sensing the large supply of latent heat energy lying over the open ocean, and not feeling as bound to form over the islands as would its African and South American counterparts, this mass of storms may move east. The bulk of the thunderstorms can shift several thousand miles east. At the height of the 1982–83 Niño, it strayed 5,000 miles from its normal location. When the largest source of energy released into the atmosphere moves one-fifth of the way around the world, things are bound to happen!

What does happen when the thunderstorms move east is that the subtropical jet stream likewise extends farther east. At times, the jet may knife into the California coast at full strength, and the incessant rains that Mark Twain considered such a feature of San Francisco's climate are at their worst. Since the subtropical jet is usually at fairly low latitudes (hence its name), many of its attendant cyclones take the southerly storm track across the Southwest. Not only does the West Coast get more storms than usual, but these storms strike farther south than usual. Even San Diego gets treated to those incessant rains—their wettest year on record, 1941, was during a Niño.

NIÑO WEATHER

The classic pattern of Niño weather was

followed in 1982–83, with the coasts of California and Oregon receiving repeated batterings from November through March. The subtropical storm track continued across the southern United States, unleashing high winds, floods and even some tornadoes along the Gulf of Mexico. Meanwhile, the northern United States was high and dry, well north of all the action; Wyoming had one of its warmest winters on record.

The arrival of spring brought no real relief from the ravages of the subtropical jet, just a change of targets. The jet performed its normal seasonal swing north, and frequent storms began dumping heavy snows on the Rockies. In May and June, melting sent rivers of water through the streets of Salt Lake City. Floods devastated the banks of many rivers of the Southwest, particularly the Colorado. Total losses from the flooding exceeded $1 billion.

The subtropical jet disappeared during the summer, as it usually does, but in a real sense the memory lingered on. The warm water of the tropical Pacific had spread up the West Coast, and the cooling effect along the coast weakened considerably. In September, the fog cleared off and Eureka, California, had its hottest day ever—a

EL NIÑO: THE VIEW FROM SPACE—These images from NOAA's Geostationary Orbiting Earth Sateiiite (GOES) graphically demonstrate the effect of el Niño on the climate of the West. This picture, taken in November 1981, shows the fairly "normal" situation before el Niño. A huge cyclone spins in the Gulf of Alaska, while clouds along the jet stream approach the coast of Oregon. Scattered tropical thunderstorms dot the lower part of the picture.

NOAA

sultry 86 degrees. At the end of September, moisture drawn from the exceptionally warm Pacific by Tropical Storm Octave led to Arizona's worst flood disaster in history.

The memories even continued into the following winter of 1983–84. The strong jet across the Pacific during 1982–83 separated the cold air to the north from the warm air to the south. The underlying ocean absorbed this temperature difference, leaving the far northern Pacific cooler than usual and the subtropical Pacific around Hawaii and vicinity warmer than usual. It takes a while for something as massive as the Pacific Ocean to change its temperature, and the temperature

pattern it picked up in 1982 and 1983 stayed into 1984. Just as the ocean reacts to the warmth or cold of the overlying air, the atmosphere tries to adjust to the temperature of the ocean below. The result? Unusually cold air over the far northern Pacific, and still warm around Hawaii—all due to the ocean retaining the temperature pattern of the previous year. This helped keep the jet strong long after all vestiges of el Niño disappeared from the tropics, and the winter of 1983–84 repeated much of the storminess of its predecessor. Like a conscience, the ocean reminds the atmosphere of its earlier transgressions.

The weather patterns of 1983 were similar

ONE YEAR LATER—El Niño is in full swing. The tropical thunderstorms have mushroomed over the Niño-warmed waters, sending clouds toward southern California along an intensified subtropical jet stream. The North Pacific jet stream has weakened, but is still there. These two jet streams—the "split jet"—mean more storms for the southern tier of states and mild weather for the north.

NOAA

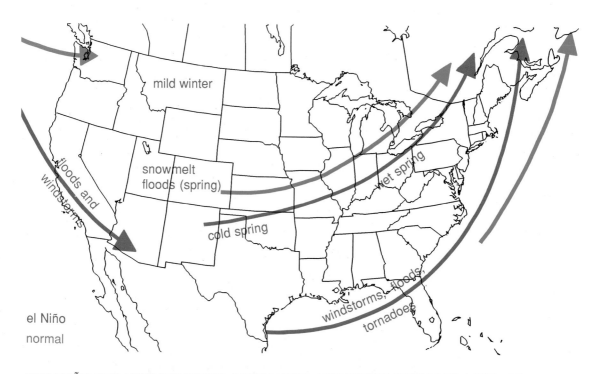

1983 NIÑO-RELATED WEATHER ANOMALIES AND THEIR STORM TRACKS—The strange weather of the 1982–83 el Niño. For the most part, it can be explained by the shifted storm tracks.

in many ways to those of 1878, 1941, 1958 and 1973, all years with large Niños. Each of these years featured stormy winters in California and mild winters across the northern states, followed by snowy springs in the Rockies and warm summers along the coast. But there were important differences among these Niño years, particularly in the timing and duration of the unusual weather. Sometimes the stormiest weather started a year early!

An explanation is needed for the drought winter of 1976–77. Although a Niño year, this winter's weather was nearly the exact opposite of that experienced during the 1982–83 and earlier Niños. Once again, the Niño off the coast of Peru was not the problem. Despite the problems around Peru, the western part of the tropical

Pacific did *not* warm up like it did in 1982. There was a slight warming, and the Indonesian thunderstorms did move east a little. But they did not move far enough east to steer the subtropical jet into the West Coast. Instead, the jet veered far to the north, crossing into the Yukon before plunging back south into the heart of Dixie. This enormous bulge in the jet, with its western ridge and eastern trough, led polar air deep into the southeast, while leaving the West completely out of the storm track. The result: persistent drought.

A NIÑO IS BORN

We have seen that thunderstorms and ocean currents in places as distant as New Guinea and Micronesia can do strange things to the weather of the West. Accordingly, a few words are in order about the tropical Pacific Ocean and why its currents fluctuate. It begins with the sun heating the ocean's top layer. Being lighter than cold water, the warm water stays on top. Most of the time, winds in the tropics blow from the east—these are the "trade winds" of sailing fame. The trade winds blow the warm water to the west, toward New Guinea, where it starts to pile up. This leaves cooler water in the eastern Pacific; this pattern of warm water to the west and cool water to the east is the normal situation in the tropical Pacific.

If the trade winds slacken, all that warm water piled up around New Guinea starts to slosh back toward Peru. Sometimes the trade winds actually reverse and blow from the *west*; this gives the pile of warm water an extra push toward Peru. When this happens, the normally cool eastern Pacific gets overrun by warm water, and el Niño is in progress. Eventually, the easterly trade winds return, and the warm water heads back toward New Guinea.

This, in simplest terms, is the cause of el Niño. The real details are not so simple. The patterns of warming and cooling differ from one Niño to the next, as do the timing, duration and

FLOODING IN SALT LAKE CITY—Excess snowmelt runoff following the wet winter of 1982–83 had to be channeled through downtown Salt Lake City, Utah.

Michael Mee, FEMA

strength of the events. One very basic question remaining for researchers is, why do the trade winds weaken? When this and other still-open questions are answered, we may be able to forecast winter weather patterns months in advance.

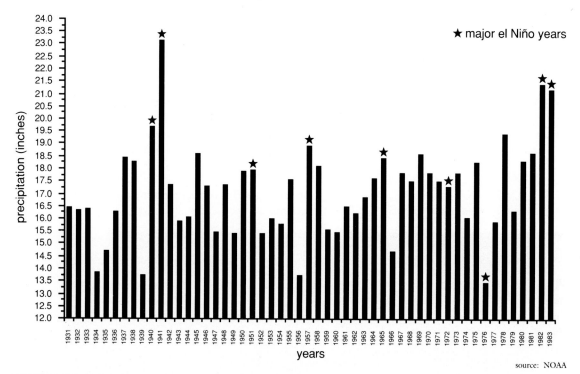

WESTERN AVERAGE PRECIPITATION, 1931–83—Fifty-three years of annual precipitation, averaged over the 11 western states, show a great deal of year-to-year variability. A striking pattern in this variability is the overall wetness of el Niño years. The wettest years of record, 1941 and 1983, were both el Niños. However, the Niño year 1976 was a severe drought across the West. Clearly, not all Niños are the same!

CHANGING CLIMATE

When asked if he thought there was anything to the climate-change business, Mark Twain is reported to have replied, "Ma'am, the weather is *always* changing!" It was his way of expressing that old maxim that the only thing that never changes is change itself. It's almost trite, but as far as the climate is concerned, it's also quite true. Whether we look at the differences from year to year or between the present and the great ice age hundreds of centuries ago, the records of western weather abound with climate changes.

First, we'd better make it clear what is meant by the word "climate." It's not the same as "weather." The latter is the day-to-day, even hour-to-hour, fluctuations of clouds, winds, temperature and the like, due to the passage of such things as thunderstorms, fronts and cyclones. Climate is the sum total of all this weather over a period of time.

There are many ways to express this thing called climate, depending on the information on hand. Lacking barometers, our ancestors used descriptive methods, recording crop successes and failures, freezes and thaws, and great deluges. This being the age of computers and weather instruments, the preferred method nowadays is to use numbers. Lots of them. There are a bewildering variety of things in the atmosphere that can be measured, and new ones are being thought up all the time. The most useful measurements for observing climate are the basics—temperature and precipitation. Long records of both are the cornerstone of climate studies; in the West they go back a century and a half. For earlier periods of time, we must rely on evidence supplied by tree rings, silt deposits in lakes, features gouged in the ground by now-gone glaciers, and even comments written by the earliest settlers.

Many folks like to think of their climate in terms of a "normal" about which the weather fluctuates. "Normal" is usually taken to mean the average temperature, or whatever, over a suitably

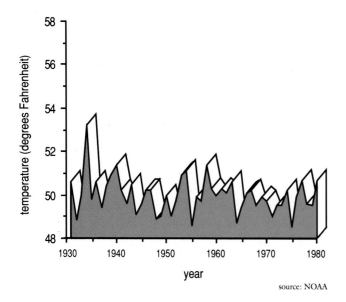

source: NOAA

WESTERN AVERAGE TEM-PERATURE, 1931–80—Fifty years of average temperatures across the 11 western states show little evidence of any long-term climate trends. What this record shows is enormous year-to-year variability that is, for the most part, still unexplained and unpredictable. The drought year 1934 was by far the warmest year on record in the West, while the cool years were 1955, 1964 and 1975.

long period of time (10 years or more). However, these averages depend on the selection of years that go into them, and, like the weather, are always changing. The truth is that there really is no exact normal. What we call normal today would have been considered quite unusual during the ice age or back in the days of the dinosaurs, or even a century ago. As the climate changes, so does our perception—and definition—of what is normal.

The shortest period of time that can be realistically called climate is a year, and the quickest kind of climate change is the variability between one year and the next. On the other end of the scale is the five-billion-year evolution of our planet's atmosphere, from a steamy soup of poisonous gases to the life-sustaining air of today. You can read about the history of the atmosphere in the first chapter; let's look at some of the things the climate has done since.

WHAT A DIFFERENCE A YEAR MAKES

For us mortals whose lives span less than a century, the most familiar and noticeable versions of climate change are the differences between one year and another. One summer may kill crops with drought, while the next drowns them with incessant rains. Some winters you may ski a trail that other years is better suited for roller skating. Winter temperatures averaged over the whole 11-state region have changed as much as 11 degrees between one year and the next, and average temperatures for entire years have fluctuated by half that. Yearly rainfall totals for the region can be twice as high in wet years as in dry ones, and even greater contrasts are seen locally. At San Francisco, yearly rainfalls have ranged from seven to 49 inches.

The year-to-year variability of climate in the West is enormous. In terms of averages, the changes are equivalent to moving from San Francisco to Seattle, or from Denver to Saint Louis. In fact, the changes are nearly as big as those that occurred during the great ice age! The distinction is, of course, that the ice ages went on for thousands of years, while the current fluctuations last a year or a season at a time. Nonetheless, we are talking about a rapid and large climate change.

The year-to-year climate fluctuations can have enormous economic consequences, in large part due to the very rapidity of the changes. Agricultural practices, for example, can adjust to the slower climate trends lasting decades or more. However, imagine moving a California citrus orchard to Seattle for a year. It wouldn't do very well, would it? That's what an extremely cold year in California can be like.

In the previous chapter we saw the marked effects of el Niño on the climate of the West. However, Niños are relatively rare, and 10 years may pass between the really big ones. What happens the other nine years? There are all sorts of possibilities; the most plausible invoke some changes in the amount of heat entering the atmosphere. Remember, the need to carry heat to the poles is what drives the weather in the first place. El Niño is an example of how huge shifts in the heat contained in the tropical ocean can affect climate.

Similar shifts of warm and cold water can take place in the North Pacific, and no doubt affect the weather of the West. When the temperature contrast of the Pacific between Alaska and Hawaii is exceptionally great, the air above

WESTERN CLIMATE: THE PAST 1,000 YEARS—Nature's record of climate is written in the growth rings of trees. Just as trees in arid climates grow more during a wet year, trees surviving near timberline grow more when their growing season is unusually warm. The widths of the rings in bristlecone pines in the White Mountains of California, averaged over 20-year intervals, tell the story of the past 10 centuries. The record reveals a warming from the ninth through the 12th centuries A.D., while the Anasazi were flourishing in the Southwest. With warmth came drought, however, and the Anasazi culture disappeared in the 1200s. We are now in another slow warming trend that began three centuries ago. Longer-term records show that the current century, along with the 1200s, are the warmest centuries since 2700 B.C.

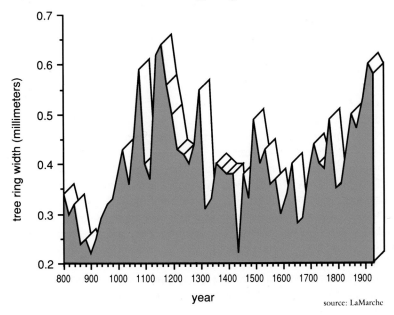

source: LaMarche

tends to take on the same contrast, strengthening the jet stream. On the other hand, a weak temperature contrast weakens the jet. If the Pacific is cold to the west and warm to the east, the jet tends to follow the boundary by veering to the north. And so on. If it has nothing better to do, the jet tends to follow the boundaries between areas of warm and cold ocean water. Sometimes it happens that way. However, from day to day the jet stream can, and does, stray far from its "expected" position.

Snow on the ground can also influence the heat balance of the atmosphere. Fresh snow is quite good at reflecting away sunlight that would otherwise heat the ground. It is also an excellent radiator of heat out to space. Furthermore, a deep, insulating blanket of the white stuff prevents the residual warmth of the ground from seeping into the atmosphere. If Canada is covered with heavy snows early in the winter, polar air masses forming over our neighbor to the north get a little colder than they would otherwise. Snow cover can affect the local climate too, particularly in the Great Basin. Cold air that pools in snow-coated valleys and basins may linger for weeks after it has left the nearby highlands. Cold pockets like this develop nearly every winter in valley locations such as the Salt Lake Valley, the basins of western Colorado, eastern Washington and Oregon, and the Snake River plain of southern Idaho. This can lead to marked temperature contrasts between the valleys and locations just 100 or 200 miles away.

Soil moisture is another possible culprit. We saw in the "Water for the West" chapter how damp soil from heavy rains can weaken the summertime ridge over the West and keep rain showers around. All these explanations have one problem in common—they don't really account for much of the variation that is actually seen from one year to the next. Every so often there is a winter or summer that looks like it suffered from one or another of these influences, but most years the weather acts in strange ways without any of these influences.

If the climate is not reacting all that often to outside forces, such as oceans and fallen snow, what is left? Can it be that it changes on its own whim? This may sound silly at first, but it is a possibility—which brings us to an elusive concept with the long name "autovariability." This has nothing to do with cars; it comes from the Greek prefix *auto*, meaning "self." Our word translates to "self-changing." So what does it mean?

We know that today's weather pattern—the positions of highs, lows, jet streams and the like—depends very much on what yesterday's pattern was. And tomorrow's will depend on what is happening today. This concept is the essence of weather forecasting. By using the laws of physics and lots of calculations, we can predict what that change will be from one day to the next. Carrying this idea further, we realize that the weather pattern, say, six months from now will be affected by today's weather map. And so will every day in between now and then. Certainly, the effect gets smaller as the interval gets longer, but it's still there. Remember, a single storm can be the most important feature of an entire winter, and a 100-mile difference in its track can make all the difference for your town. A severe winter may result from the weather "getting off on the wrong foot."

To put it in different terms, the whole sequence of fronts, cyclones and anticyclones—the exact tracks they take and the specific dates and times they pass certain points—depends on their positions on the first day of the season (or *any* day). It's like rolling a boulder down a hillside. The law of gravity and the shape of the hill guide the path of the boulder once it is rolling, but the precise point of departure—the exact position of the rock when you shoved it—decides which of the infinite possible paths the rock

120

actually takes and where it finally lands.

Unfortunately, our knowledge of the atmosphere isn't complete enough to allow forecasts to be made very far in advance. It's tough enough going five days into the future, much less six months. In effect, we know neither the shape of the hill nor the initial position of the rock accurately enough to do better. This leaves us with the prospect that seasonal forecasts, those predictions of general weather patterns made a month to a year in advance, may never be as correct as we would like.

Again, this "autovariability" is an elusive concept and difficult to understand. It's not as easy to deal with as some clear-cut, identifiable cause such as ocean temperatures or volcanoes. But it's there, and because of it, don't expect perfect long-range weather forecasts any time soon.

SOMETHING'S IN THE AIR

Now we get to some of the things that are more commonly thought of as "climate change," such as the effects of carbon dioxide, volcanoes and, yes, even nuclear bombs. Compared to something as slippery as autovariability, the influences of dust and gases in the air are refreshingly simple to grasp. The theory behind these atmospheric pollutants, some natural and some man-made, rests on how they foul up the radiation balance of the globe. The earth gains heat energy from sunlight striking the surface of the planet and loses it by infrared radiation back out to space.

The current temperature of the earth, averaged worldwide, results from the balance between heat gains and heat losses. Anything that changes the amount of incoming light or outgoing radiation, or both, should change the temperature of the earth. Yes, it's just like your checkbook—if you earn less, or spend more, or (especially) both, you go broke.

The way for planet Earth to earn less is to block out the sunshine. Volcanoes spew tremendous volumes of geologic gunk into the atmosphere, and these clouds can certainly blot out the sun. The ash cloud from Mount Saint Helens' cataclysmic eruption of May 18, 1980, kept towns from Yakima, Washington, to western Montana in the gloom for one or two memorable days. And these places did cool off from the lack of sunshine —afternoon temperatures were 10 degrees or more lower than they should have been. However, this was a local and short-term effect and qualifies more as a bizarre form of weather than as climate change.

To affect the climate, volcanic ash has to get high into the atmosphere and stay there for several years. Volcanic ash itself—those little particles of rock that gritted everyone's eyes after Mount Saint Helens—doesn't stay up all that long. Usually most of it falls from the sky after a few days and never has a chance to foul the climate. What does stay up are the clouds of sulfur dioxide gas shot into the sky from erupting volcanoes. Sulfur dioxide is the same noxious gas that comes from coal-burning factories and power plants. Saint Helens blew a half-million tons of the stuff into the atmosphere, equivalent to about five months' worth of emissions from all the industrial sources in the West.

Mount Saint Helens had relatively little sulfur, however, and much more of the stuff went skyward when the Mexican volcano El Chichon blew in April 1982. The 1,000-degree heat of the gases combined with their muzzle velocities out the volcano's throat sent tremendous amounts of sulfur dioxide into the stratosphere. The stratosphere is that part of the atmosphere lying between about eight and 30 miles up. It is a pretty quiet place, and once gases get there, they stay for months or years.

As a gas, sulfur dioxide is transparent, but in the stratosphere it combines with water vapor to

121

Winston Scott

MOUNT SAINT HELENS—The eruptions of Mount Saint Helens in 1980, such as this one on July 22, threw millions of tons of ash and gases into the atmosphere, darkening the sky across much of the United States for several days. The long-term effect of these eruptions on climate, if any, were not measurable.

Richard A. Keen

VOLCANIC TWILIGHT—Two years after Mount Saint Helens, the Mexican volcano El Chichon sent massive clouds of sulfurous gases into the stratosphere. These gases remained for years, causing brilliant lavender sunsets around the world.

form little droplets of sulfuric acid—the same corrosive material that forms the clouds shrouding the planet Venus. Fortunately, Earth's sulfuric acid clouds never get as dense as Venus', although after El Chichon they were thick enough to block out 2 to 5 percent of the sun's rays. Similarly thick clouds formed after the eruptions of Krakatoa in Indonesia (1883), Santa Maria in Guatemala (1902), Katmai in Alaska (1912), and Agung, also in Indonesia (1963). In 1815, Indonesia was also the home of the most recent really big blow, when Tambora exploded with 50 times the force of Mount Saint Helens.

It's not hard to see these clouds of sulfuric acid if you look at the right time. At 15 miles up, they catch the last rays of sunlight long after the ground has slipped into darkness, resulting in brilliant lavender twilights about 20 minutes before sunrise and after sunset. Spectacular twilights went on for several years following all the big eruptions of the past century.

Theoretically, trimming the amount of sunlight by 2 or 3 percent should cool the surface of the earth. The amount of solar energy reaching the West also drops off by 3 percent every two days during autumn, due to the lowering sun angle. Over these two days, the West's average temperature normally cools about three-quarters of a

123

degree, so we might expect a similar cooling following a big eruption. The volcanic cooling should last as long as the volcanic cloud—one or two years.

It's difficult to say whether these volcanoes really did cool the climate. The worldwide average temperature actually rose a bit following 1982's El Chichon eruption. This was, however, the same time el Niño was warming the Pacific Ocean. The global climate cooled about half a degree around the times of the Agung, Santa Maria and Krakatoa eruptions, but some of these cooling trends seemed to start *before* the eruption! The detonation of Tambora was followed by all sorts of anecdotes about "1816, the Year without a Summer" in Europe and New England. However, some of the scanty temperature records of the time—including Thomas Jefferson's readings from Virginia—fail to show any unusual cooling that year.

Climate data for the West are equally noncommittal. There are no western weather records for 1816, and any changes following the more recent eruptions are mere fractions of a degree. The court is still out on this one. Probably the only way we'll ever know how much volcanoes can alter climate is to have another titanic eruption like Tambora or the 4,400 B.C. explosion of Mount Mazama (now Crater Lake, Oregon). It will be an expensive lesson to learn. When it went, Tambora took 92,000 souls with it.

Believe it or not, we humans have devised an even more expensive way to find out what dust in the atmosphere does to climate. The trick is to have a nuclear war! The general idea is the same as volcanoes, except that soot, instead of sulfuric acid, from evaporated cities and forests is blown into the atmosphere. There's an ongoing debate as to how much soot would be created, how high it would go into the atmosphere, how long it would stay there, and what its effects on the climate would be. Some say a massive cooling, the "nu-clear winter," would result from blockage of sunlight; however, there's a counterpoint that there would instead be a "nuclear summer," since all that dust would also let less infrared radiation go back out. It's still just a theory, and one we never want to test.

We've found ways to keep the sunlight out. How about keeping the earth's heat in? The way to do this, of course, is to block the escape of infrared radiation. Several trace constituents of the air, such as water vapor and methane, are quite effective at doing this. Another gas that works well is carbon dioxide. Being transparent, these gases let sunlight in; however, they do not let infrared radiation out. It works just like the glass covering of a greenhouse, and so the idea has been coined the "greenhouse effect." We know the greenhouse effect works; just look at Venus, whose carbon dioxide atmosphere has raised its air temperature to 850 degrees.

Carbon dioxide is of particular interest. Its amount in the atmosphere has been increasing over the past century because of the burning of coal and oil. Burning wood doesn't really add carbon dioxide to the air; it merely returns the carbon dioxide taken from the air by the trees when they were growing. Burning coal and oil releases carbon dioxide that was removed hundreds of millions of years ago, when the earth's atmosphere was full of the stuff.

So far we humans have burned up about 10 cubic miles of oil, raising the amount of carbon dioxide in the atmosphere about 15 percent above its natural content of a century ago. Theoretically, this amount of additional carbon dioxide might warm the earth a fraction of a degree. Like the effect of volcanoes, any warming that may have actually happened is still too small to show up in the temperature record. A fraction of a degree is awfully small compared to the larger effects of el Niño and autovariation. However, at the rate we're burning coal and oil, it may not be long. . .

DROUGHT AND DELUGE

There will come seven years of great plenty throughout all the land of Egypt, but after them seven years of famine.

—Genesis 41:30

Cycles of rain and drought are nothing new. They plagued the pharaohs of Egypt in the times of Genesis, and today they plague the presidents and prime ministers of nations from Africa to North America. Our county had its share of problems during the great drought that scorched the central states during the 1930s. The drought area became known as the "Dust Bowl," and, curiously, it lasted seven years—from 1933 to 1940. The Dust Bowl combined with the Great Depression to make for the hardest times this country had seen since the Civil War. Many of the troubles were epitomized in John Steinbeck's 1939 novel, *The Grapes of Wrath*, describing the exodus of drought-ruined Oklahomans westward to California.

The heart of the Dust Bowl extended to the eastern plains of Colorado and New Mexico, where huge volumes of dry topsoil were lifted from the ground and blown into the Atlantic by late winter cyclones. Farther west the drought was much less severe, but dry conditions did occasionally follow the Oklahomans into California.

Drought returned to the High Plains and mountain states in the mid-1950s. This time the conditions were not as severe as they were in the Dust Bowl, and with the agricultural lessons learned from that experience, crop yields stayed above the disaster level. In the mid-1970s drought again struck the West. Although this one was less severe in the High Plains, in February 1976 a dust storm reminiscent of the 1930s carried Colorado soil to the East Coast. It was a memorable dry spell on the West Coast, however.

A glance at the dates of the great droughts shows what looks like a 20-year cycle. Indeed, from the mid-1850s up to the mid-1970s, drought has struck the plains states with remarkable regularity. The similarity between this 20-year cycle and a 22-year cycle of storm activity on the surface of the sun, known as "sunspots," has led to speculation that sunspots may, in some way, cause the drought. However, this theory is very controversial. One problem is that after six drought cycles of 20 years each, the 22-year sunspot cycle is 12 years out of sync! Some recent droughts have occurred near the peak of the sunspot cycle, but in the 1800s they preferred the dips. Furthermore, the droughts don't always strike the same parts of the plains. While the evidence for a connection between sunspots and drought may appear a bit shaky, we shouldn't write off the theory completely. Future data may show that sunspots (or volcanoes or carbon dioxide) are indeed affecting the climate! Farther west, in the Rockies and on the Pacific coast, the 20-year drought cycle doesn't show up at all. Wet and dry spells appear to fluctuate more rapidly, and may—in part—come from the ons and offs of el Niño.

Terrible as the Dust Bowl was, history hints at a much more catastrophic drought seven centuries ago. Although there were no meteorologists around at the time to make precise readings of the rainfall, trees did the job just as well. Trees grow more slowly when there's a drought, so their annual rings are thinner. The record of tree rings from that era indicates that a severe drought struck the central and western United States in the late 1200s. This drought lasted 38 years—five times the Dust Bowl years! A drought of this magnitude must have had a tremendous impact on the residents of the West, and apparently it did. Many of the pueblo cities of the Southwest, including Mesa Verde, were suddenly abandoned about the same time. The people and their culture disappeared. To tribes who later settled the area, they became known as the Anasazi —the ancient ones. Did the drought drive them away? If it did, it is a lesson we shouldn't ignore.

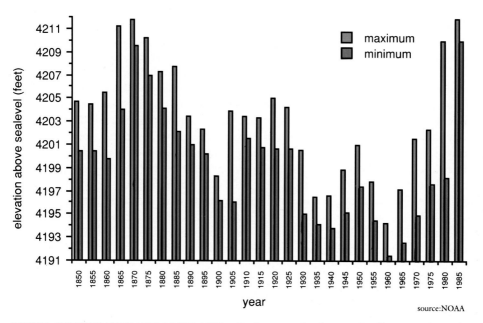

source:NOAA

LEVEL OF THE GREAT SALT LAKE—Perhaps no single record tells more about the recent climate of the West than the changing level of the Great Salt Lake. Located directly in the middle of the West and draining four of its states, the lake acts as a huge natural rain gauge for measuring western precipitation. Decreasing lake levels in the 1870s, 1890s, 1930s and 1950s show hints of a 20-year drought cycle, while wet weather in the 1860s sent the lake rising to 4,211.8 feet above sea level in 1873. Recent heavy precipitation—including some caused by the 1982–83 el Niño—has dramatically raised the lake level. In April 1987, the level reached 4,211.85 feet—an inch above the 1873 record!

THE GREAT ICE AGE

Now it's time to talk about the grandaddy of all climate changes—the great ice age. To anyone standing on a Chicago street corner on a sweaty July afternoon, it would seem absolutely inconceivable that just 18,000 years earlier (a mere blink of an eye, geologically speaking), that same spot was buried under two miles of solid ice. In January, however, it might seem a little more believable. Sheets of glacier ice once extended as far south as New York City and Saint Louis; the Ohio and Missouri rivers form the approximate boundaries of the ice sheet. Other ice sheets covered Antarctica and parts of Europe, from Britain to the Soviet Union.

The greatest ice sheet of all time was the one that initially formed around Hudson's Bay, and spread over the northeastern half of the continent. This mass of ice did not make it very far into what is now the West and just barely entered northern Montana. However, mountain glaciers in the northern Rockies and Cascades grew and coalesced into a smaller ice sheet that filled Puget Sound and later merged with the larger continental ice sheet.

Although most of the West remained free of the ice sheet, there is a great deal of evidence that the regional climate was vastly different from today's. Mountain glaciers in the southern Rockies and Sierras grew for miles beyond their present locations, and what are now alkali flats and salt lakes in the Great Basin were then huge freshwater lakes 100 or more miles across. The largest of these was Lake Bonneville, which extended for 300 miles across Nevada, Utah and Idaho. At times, glaciers from the Wasatch Mountains may have flowed into this inland sea. All this points to a much cooler and wetter climate than we have now.

Not only was it colder, it was also windier during the ice age. Geologic relics from the old days, such as wind-scoured rocks found along the eastern slopes of the Rockies, suggest that the prevailing westerlies blew stronger back then. We might expect this, since the Arctic cooled off much more than did the tropics, increasing the equator-to-pole temperature difference. This, in turn, would strengthen the jet stream. Thus, western weather during the ice age may have seen continuous storms blasting the West Coast and dumping snow in the mountains, with incessant chinook winds howling east of the Rockies—even during summer. It's speculative, of course, but plausible.

The climate of the West has warmed and dried considerably since the ice age. Lake Bonneville is now gone; in its place are the Great Salt Lake and the seasonally wet Bonneville salt flats. At its peak, Lake Bonneville was more than 1,000 feet deeper than the present Great Salt Lake. Compared to recent fluctuations of the level of the Great Salt Lake, we can see how incredibly different the climate of the ice age truly was.

When did all this happen? Apparently, there have been a whole series of ice ages, beginning several million years ago with the growth of the Antarctic ice sheet. For the past 700,000 years, ice sheets have grown and melted in the Northern Hemisphere at 100,000-year intervals and lasted several tens of thousands of years each. Between each of these ice ages, the climate was much like it is today. The last ice age began 75,000 years ago, and the ice sheets grew to their maximum extent 18,000 years ago. Four thousand years later, around 12,000 B.C., the world's climate suddenly warmed up to near present temperatures. It took another 5,000 years to melt the huge mass of ice over North America, and the last patch disappeared from northern Quebec around 5,000 B.C. The ice age is still continuing in Antarctica and Greenland, and remnant ice caps are still found on some of the islands of northeastern Canada.

There are many theories about the cause of the ice ages, and some of them may actually be true. One plausible theory was put forth in 1920 by a Yugoslavian named Milankovitch and concerns the wobbles of the earth and its orbit. Like a spinning top, the axis of Earth's poles wobbles around and changes its tilt in cycles of 26,000 and 40,000 years, while the shape of the earth's orbit around the sun changes over several longer cycles—up to 100,000 years. The actual size of the orbit never changes, and, over an entire year, the total amount of sunlight reaching the earth never changes. Thanks to these wobbles, however, the solar heating of the different hemispheres during different seasons varies by up to 5 or 10 percent over thousands of years. If the Northern Hemisphere gets less sunlight during the summer, that season is cooler. As it is now, the winter snows that cover the far north barely melt before the summer is over; if the summers were 5 degrees cooler, the snow might not melt at all from some places. That means the snow would build up year after year. Once the snow cover gets thick enough, it can increase the cooling by reflecting sunlight back into space. Keep that up, and in a few thousand years there's an ice sheet.

The Milankovitch theory predicts several cycles that might affect the buildup of snow, and some of them do fit the cycles of the ice ages. The theory also predicts that the past ice age was not the last one. So if you thought the winter of 1949 was bad, just wait until 10,949!

WATCHING THE WEATHER

"I wish you had a thermometer. Mr. Madison of the college and myself are keeping observations for a comparison of climate." These words were written in 1784 by Thomas Jefferson to the other Mr. Madison, James Madison, the fourth president of the United States. Jefferson had been keeping weather records since 1777 and continued to do so for 50 years. He logged his valuable account of early American weather into his *Weather Memorandum Book*, now on file at the Library of Congress and the Massachusetts Historical Society. James Madison took Jefferson's advice and soon began taking thermometer readings at his home. For his last 33 years, George Washington kept an "Account of the Weather." Thus, the first, third and fourth presidents of the United States knew the joys (and importance) of watching the weather. They understood that the best way to learn about something is to participate in it, and to participate in the changing weather means to observe it closely.

Humans haven't always known what makes the weather work. That knowledge came over the centuries by paying close attention to the passages of clouds, wind and rain, and to the vagaries of temperature, humidity and pressure, and finally fitting all the observations together until they made sense. There is no need for you to recreate these early days of discovery, of course, but if you watch the skies carefully enough and long enough you find that you understand the weather in a way no book can teach you.

Yogi Berra understood these great truths, and even said so: "You can observe a lot by just watching." Baseball managers aren't meteorologists, but in many ways, weather is like baseball. The ever-changing action is fascinating and fun to watch. The more you know about what you're watching, the more rewarding it is. If you know more than the umpire (or any other expert), it's even more fun yet! How do you keep score? By taking records of temperature, rain, snow or whatever else suits your fancy. (Whether your daily temperatures look more like baseball, football or basketball scores depends on where you live.)

Observing the weather is an activity that can be done at nearly any level of dedication and expense. On one extreme, you can simply watch the clouds and what falls out of them and note what kind of weather occurs each day. You don't need to spend a cent on weather instruments since you already have the most remarkable weather instrument ever devised—a pair of eyes. On the other extreme, you can invest thousands of dollars in automatic recording weather devices, perhaps even hooked up to a home computer. Most of you will fall somewhere in between these two extremes; you may want to watch the weather with the help of a thermometer and a rain gauge. After all, it is in some strange way satisfying to be able to put numbers to things. To measure something is to understand it a little bit better, and being able to compare one weather event with another makes both all the more familiar. Weather is indeed like baseball; we all want to know how yesterday's snowfall stands in the all-time ratings.

Now that you're convinced you want to set up a small, inexpensive weather station, what instruments should you get? Should you buy them or make them yourself? Where do you put them and how often do you read them? And what do you do with your records? Of course, your choices are individual ones. But perhaps this veteran weather buff can give you some hints and cautions to ease you into a fascinating hobby that only gets more captivating with time.

No matter how big or small your weather station is, or even if you're using no instruments at all, keep in mind that some form of record keeping is essential. Memories change with time, with big storms getting bigger and small storms smaller. Most days disappear entirely. Twenty years from now, you probably won't remember what today's weather was. Write it down, and two decades from

now your notebook will be one of your prized possessions. If you only have a dollar to spare, invest in a notebook. Roomy calendars and appointment books work just fine, as do business ledgers and cash books.

Now for the instruments. It pays to be somewhat picky in your choice of instruments and to be careful about how and where you set them up. You want your records to be as accurate as possible. Not only is it a matter of personal pride, accurate records can be of use and interest to others. The following suggestions should help you get the best weather records for your dollar.

TEMPERATURE

The most basic instrument—and one that most people already own—is a thermometer. However, most household thermometers don't give very accurate readings. This is not to say the thermometers themselves are inaccurate, but rather they are situated where their readings are not representative of real weather. A window-mounted thermometer picks up some heat from the house and may read 5 or 10 degrees higher than the actual outside air temperature. A thermometer sitting in the sun might read 50 degrees too high. While temperatures "in the sun" may mean *something*, it is difficult to say what. Consider that the reading of a thermometer set in the sun is affected by wind speed, angle of the sun, where the thermometer is mounted, type and brand of thermometer and, last and least, actual air temperature. In other words, you don't learn much from thermometer readings taken in the sun. That is why meteorologists around the world have chosen to take readings "in the shade." If your thermometer readings are taken the same way, they can be compared directly with temperatures from, say, Bangkok or Siberia, or even from a spacecraft on the surface of Mars.

To a meteorologist, "in the shade" doesn't necessarily mean under a tree. In fact, it has been found that thick groves of trees generate a local "microclimate" that is substantially cooler than surrounding areas. Good locations for thermometers are on the north sides of buildings or under porch roofs, although even here reflected sunlight can affect the readings by a few degrees. The best location for a thermometer, and the one chosen as an international standard by the World Meteorological Organization, is in a white louvered box at eye level above open ground, as far as possible from trees, walls, buildings or other obstructions. While setting the thermometer out in the open may seem to contradict the "in the shade" requirement, the louvered box (called an "instrument shelter" in the business) actually provides more effective shade than any tree or wall, while still allowing free flow of air around the thermometer. Anybody willing to invest $30 and a day's work can build a suitable box. A design that has served me well for many years is described in Appendix 2. Making one would be an easy project for a novice do-it-yourselfer.

Whether or not you make a box, weather watching is a lot easier if your thermometer can tell you what the temperature *has been*, as well as what it is now. The simplest (and cheapest) way to do this is to get a maximum-minimum thermometer, called simply a "max-min" in the trade. This kind of thermometer indicates the highest and lowest temperatures that have occurred since you last checked it. Simply read the high, low and current temperatures, reset the high and low indicators, and come back the next day. The best max-min thermometers are the liquid-in-glass type. Good ones made by Taylor and Airguide are sold in hardware and department stores for $20 to $30. There are cheaper max-min thermometers of the coiled-metallic-spring type, but their metal springs have a tendency to shift and stretch with time and cause the temperature readings to lose accuracy. For $100 and up you can get chart recording or computerized memory thermometers that give you the same information and more. However, these expensive gizmos are complicated and can sometimes break down. For example, computerized thermometers are very susceptible to lightning. The basic max-min thermometer is a glass tube with some mercury and alcohol in it, and about the only way to make it malfunction is to take it out of its box and hit it.

Temperatures are best recorded on a daily basis. With an ordinary thermometer, readings made at the same time every day, say 7 a.m. or 6 p.m., give a good account of day-to-day excursions of weather. By taking 24 readings—one every hour on the hour—and averaging them, you get the average temperature for the day. That's a lot of work. Fortunately there's a better way. Nearly two centuries ago, weather watchers found that the average of readings taken at sunrise, 2 p.m. and 9 p.m. came reasonably close to the actual average temperature for the day. Since the max-min thermometer was invented in the last century, an even easier way to figure average temperature was discovered: simply average the high and low temperatures for the day! All you have to do is read your max-min thermometer once a day and you've got the day's average. Read it each day for a month and you can figure the average for the month, and so on.

Even with a max-min, there are good times and not-so-good times to read your thermometer. For example, if you read your max-min at 6 a.m., the temperature is near the minimum for the day. If that minimum was, say, 48 degrees, the temperature when you take the reading may be 50. When you reset the thermometer, it remembers that 50-degree temperature. Now imagine that the weather warms up over the next 24 hours (probability says this happens about half the time!), and the next morning's minimum is 60 degrees. When you take the reading, however, the thermometer will show a low of 50, left over from the previous, cooler morning. This happens often

Richard A. Keen

RAINBOW—A summer shower ends with this brilliant rainbow in northern Wyoming.

enough that your computed averages can be as much as 2 degrees lower than the actual average temperature. The opposite problem occurs if you take your daily readings in the afternoon—your averages will be 1 or 2 degrees too high. The best time to read a max-min thermometer is when the temperature is normally midway between the high and low. Studies of this problem have shown that these times are about 9 a.m. or 10 p.m. The National Weather Service has chosen the calendar day—midnight max-min readings—for their weather stations. Your choice of a time depends on your lifestyle. One important thing to remember is that the key to good and useful records is consistency. Pick a time that you can stick with over the long run. There are a lot of false "climate changes" in weather records the world over that are really changes in weather station locations or observa-tion procedures. Try not to change your horse in midstream, and your records will be all the better for it.

HUMIDITY

Much of what we call "weather"—clouds, rain, snow, frost, fog and so on—is caused by moisture in the atmosphere. This is why humidity is so important. Humidity is also difficult to meas-ure accurately. Most common humidity indica-tors, or "hygrometers," rely on the fact that some substances—such as hair and paper—expand when they absorb moisture from the air. Unfortu-nately, these substances attract dust and bugs and also tend to stretch over time. As a result, the humidity readings are not a whole lot more accu-rate than noting how clammy your skin feels when you're outside.

131

Another problem with hair or paper hygrometers is that they measure *relative* humidity, which is the amount of moisture in the air expressed as a percentage of the amount the air could hold at that temperature. However, the really interesting thing to measure is the *absolute* humidity—the actual amount of moisture in the air. Meteorologists prefer to express absolute humidity in terms of the "dew point." If you take outside air and cool it, the dew point is the temperature at which the water vapor in the air condenses. In other words, when night cooling drops ground temperature to the dew point, dew forms. Higher dew points mean more moisture in the air.

The most economical way to measure dew point is with a "wet-bulb hygrometer," available for $30 or $40. If you don't like the price, it's easy to make your own for about $5. This little gadget consists of two thermometers mounted side by side. One thermometer has a wet cotton wick attached to its bulb; the other end of the wick is dipped in a small reservoir of water. The dry thermometer reads the ordinary air temperature, or "dry-bulb" temperature, while the wet thermometer gives, not surprisingly, the "wet-bulb" temperature. Because of evaporative cooling, the wet-bulb reading is always lower than the dry-bulb reading. Using a small chart supplied with the hygrometer, you can convert wet- and dry-bulb temperatures into dew points.

Another version of the wet-bulb hygrometer is the "sling psychrometer." This also has dry- and wet-bulb thermometers mounted side by side, with the whole thing attached to a hinge or a chain. Soak the wet-bulb wick, twirl the thermometers around (to get better ventilation), and read the dry- and wet-bulb temperatures. With either version, the situation gets complicated when the wet-bulb temperature is below freezing. Ice evaporates or, more properly, sublimates more slowly than liquid water, so when the wet wick freezes, the ice-bulb temperature reads higher for the same dew point temperature. There are separate charts for converting ice-bulb readings to dew points.

Watch the dew point closely, especially during summer, and you'll see how the moisture content of the air changes with different air masses. In desert areas, dew points rise with the onset of the summer monsoon. If it gets above 60 degrees, thunderstorms may be in the offing. In coastal regions, the dew point changes sharply as winds switch to and from the ocean. Eventually, you'll find that the dew point is a handy number to know when making your own forecasts.

RAINFALL

Measuring rainfall is the epitome of simplicity. "One inch of rain" from a storm simply means that the ground is covered with one inch of water (assuming none of the rain runs off or soaks into the ground). Rain is a lot easier to measure than temperature and in some ways is more interesting. Rainfall rates can vary dramatically over distances of a few miles or less. Unless you live close to a weather station, you really need your own measurements to know how much rain you've had.

Official-type rain gauges cost several hundred dollars, but fortunately you can buy small gauges that are nearly as accurate for about $5. My favorite is the plastic "wedge" made by Tru-Chek, but others of the same genre are just as good. To set up your rain gauge, mount it on a fencepost, clothes pole, or (for you apartment dwellers) up on the roof. About the only rule to remember is not to place the gauge near or under anything that might interfere with the rainfall. The rule of thumb is to keep the gauge as far away from trees, buildings and the like as the obstructions are high. Read the rainfall at the end of each storm (don't wait too long, or some rain might evaporate) or, if you like, make daily readings when you check your thermometer. Don't forget to empty the gauge each time and, above all, remember to write down the amount before you forget it.

SNOWFALL

Although snowfall is one of the trickiest of weather phenomena to measure, its measuring instrument is certainly the simplest and most commonplace of all weather instruments. Believe it or not, even the pros at the National Weather Service use an ordinary ruler (or yardstick) to measure snowfall. The difficulties arise because of what happens to snow after it falls. If it's warm, snow melts as it hits the ground; if it's windy, snow drifts; and if snow accumulates over 4 to 6 inches, it settles under its own weight. Because of the varied fates that can befall grounded snowflakes, snow depth on the ground sometimes decreases even as the snow continues to fall. So really, there are two things to measure. One is snow depth; the other is snowfall.

To measure snow depth, simply go out and stick your ruler into the snow in several places and take an average depth. If there has been a lot of drifting, take more measurements until you feel confident you've got a good average. This technique is simple and straightforward. Snowfall, on the other hand, requires an attempt to measure the amount of snow that would have accumulated had there been no melting, drifting or settling. To do this, measure snow depth on a surface with relatively little melting—wooden decks and picnic tables are favorites. Some people prefer to use a special "snow board," a white painted piece of wood set on the ground. Measure the snow every few hours, or whenever 4 to 6 inches accumulate, and write down your measurement. Clear off the table or set the snow board back on top of the snow. Let the snow start accumulating again and take another measurement in a few hours. This eliminates the settling problem. Your snowfall total is the sum of the individual measurements. It's not really all that difficult, once you have the procedure down.

Measurements of snow depth and snowfall often differ, particularly in bigger storms. Snowfall is always the larger of the two. The National Weather Service measures both, but uses *snowfall* for the official record of a storm. If you live near an official weather station and wonder why they always have more snow than you, it may be because you're comparing their snowfall with your snow depth.

WIND

Wind can be as tricky to measure as snowfall and a lot more expensive. There are an incredible variety of wind gauges—or anemometers—available on the market, most of them of the familiar three-rotating-cup variety. For about $100 you can buy one that tells you wind speed; for twice that they throw in wind direction. For $300 there are anemometers that recall the peak gust since you last reset it, sort of like a max-min thermometer. For the same amount of money you can get one that gives a continuous chart recording of wind speed. On the opposite end of the price scale is an elegant little device that measures wind speed by using the pressure of the wind to push a red fluid up a tube. Made by Dwyer, this gauge sells for about $40. Whichever design of wind gauge you may decide to buy, remember to place it high enough that trees and buildings don't interfere with its readings. Above all, ground it well to protect it from lightning.

Anemometers, with their rapidly fluctuating dials, are fun to watch. However, you don't absolutely need to buy one to measure wind speed. You may very well have one or more anemometers growing in your backyard, because trees are fairly accurate indicators of wind speed. The idea of watching the effects of wind to estimate wind speed was first formalized in 1806 by Sir Francis Beaufort of the British Navy. Beaufort tabulated descriptions of the state of the sea—wave heights, roughness, whitecaps and the like—correspond-

Richard A. Keen

DOWNPOUR—Gray sheets of rain pour from the base of a summer thunderstorm.

ing to different wind speeds, for use by mariners. Later, Beaufort devised a similar table for land use. More recently, Theodore Fujita of the University of Chicago expanded the scale to the extremely high wind speeds found in tornadoes and severe hurricanes. These wind scales are described in Appendix 3 and you can use them to make surprisingly good estimates of wind speed.

The diehard do-it-yourselfer might want to make his or her own anemometer. It's not impossible, as the basic concept is really rather elementary. The wind pushes on the three round cups and makes them spin about the axis. This in turn rotates a small electric generator which creates a voltage (proportional to wind speed) which moves the needle on a voltmeter. A small direct-current

motor from a hobby shop works just as fine as a generator, and a milliammeter makes a great readout dial. Making the cups and attaching them securely to the motor can be tricky, as can waterproofing the motor from the rain. An effective way of calibrating the device is to drive a car at specific and constant speeds while holding the cup mechanism out the window and comparing the current readout with the speedometer. Enlist a friend to help you with the driving.

The not-quite-so-diehard do-it-yourselfer may prefer to build a wind vane, which simply indicates wind direction. The standard design is essentially a large-tailed arrow balanced on a pivot, so the arrow always points into the wind. This can be as simple as a plywood arrow that

134

pivots on a nail. Let your imagination run free on this project. When you record wind directions, though, remember that meteorologists always talk about the direction the wind blows *from*.

BAROMETRIC PRESSURE

The most popular instrument for do-it-yourself weather forecasters is undoubtedly the barometer. Most weather books contain tables describing tomorrow's weather based on today's barometric readings. Unfortunately, these tables don't always work in the West. We all know that falling pressures mean stormy weather and rising barometers mean clearing, right? Not necessarily so! There are places in the western states where the heaviest storms begin as the pressure starts to rise and places where falling pressures bring sunny weather. You're on your own here. Buy a barometer—decent ones are $20 and up—and read it once or twice a day (preferably at the same time each day). Don't forget to write your readings down, along with the rest of your observations.

WHAT TO DO WITH YOUR WEATHER RECORDS

When you first start taking weather records, the satisfaction you gain will be a personal one. Each time you watch, measure and record some weather phenomenon, you notice details about the event that would have otherwise escaped you. You may certainly have some fun gathering your own statistics. After a while, you'll have enough statistics to start tinkering with them. Plotting temperatures, pressures, rainfall and snow on graph paper is an engaging way to display your data, and gives you a real feel for how the weather varies from day to day. This job is easier if you have access to a computer with spreadsheet and graphics software, but plotting it by hand lets you appreciate the data more. After all, the computer doesn't care what these numbers mean! When you have several years of records, you'll be able to calculate "normal" temperatures, rainfall and snowfall for, say, April, and have a record of all-time highs and lows.

Don't forget that there are others who also like statistics. They may be interested in yours. Just look at a newspaper. Check the sports page or the business section or even the weather column, and what do you see—statistics! People love statistics, so don't be shy about sharing yours with them. How you go about doing this depends on where you live. If you're in a city or near an official weather station, you don't want to compete with the official records. However, the public, media and even meteorologists are interested in how the weather varies across town, especially during storms. Let your newspaper, radio or television station, or National Weather Service office know you're keeping records. You may be pleased to find they're interested in your reports.

If you live in a small town or in the country away from an official weather station, you have a ready-made audience. Your local newspaper may be delighted to get your weather reports on a regular basis. I live in a small community in the Colorado Rockies and write a monthly weather column for a local newsletter. I've heard that some residents actually clip and save my column!

The greatest need for weather observers is in isolated regions of the West. The National Weather Service has organized a nationwide network of volunteer weather observers in a program called the "Cooperative Observer Network." The role of these cooperative observers is to fill in the large gaps between National Weather Service stations, which are usually located at major airports. Ideally, cooperative observers are spaced about 20 miles apart, with 100 or 200 such observers in each state. If you live more than 20 miles from the nearest cooperative station, the National Weather Service may want *you* as an observer! If you become a cooperative observer, the weather service will set you up with official U.S. Government-design weather equipment. Your responsibilities include taking daily observations on a consistent basis and filling out and sending in a monthly weather summary. The satisfaction comes in seeing your data published in the government publication, *Climatological Data*, and knowing that thousands of subscribers across the country are reading and using your data.

If you like kids, this is your big chance to enrich their lives. Schools and scout groups are always looking for projects and field trips.

Wouldn't you have liked to visit a weather station when you were a kid?

Sharing your interest in the weather can be rewarding. Eventually, however, your greatest satisfaction will probably come from your own growing understanding of the way weather works. Even though weather can cause personal hardship—from leaking roofs and snow-slick roads to planes missed because of fog—following it over the years will lead to a familiarity that breeds respect rather than contempt.

WEATHER MAPS

You can look out the window to see what the local weather is doing, but you need a broader perspective before the weather you see makes any real sense. After learning this in the 19th century, meteorologists developed the weather map. Since then they have devised internationally accepted ways of mapping weather. Weather maps used by meteorologists are loaded with information expressed by hundreds of symbols and numbers. There's no need to learn all the fine details of a weather map, but you may wish to pick up some of the basics. The aim of this chapter is to point out the most important features of the weather map and to show an example of a weather map from a notable day.

Unless you're a sailor or a pilot, you probably don't have many chances to see a real weather map. There are weather maps that appear every day in newspapers and on television but, unfortunately, these media maps are often simplified to the point that they tell very little about what's really happening. You can, however, subscribe to a booklet of daily maps published weekly by the National Weather Service. (See the next chapter for the address.)

The first thing you notice when you look at a weather map is a lot of lines. Many of these are "isobars," or lines of equal barometric pressure. They're very much like the contour lines on topographic maps. Winds tend to blow along the isobars, clockwise around highs and counterclockwise around lows, with a further tendency to angle in toward low pressure. The closer the isobars are to each other, the stronger the pressure gradient and the faster the winds. The many isobars around an intense cyclone sometimes look like a bull's-eye.

On weather maps, isobars are labeled in "millibars" of pressure. While your barometer probably reads pressure in inches of mercury, meteorologists generally use millibars to measure pressure. The average atmospheric pressure

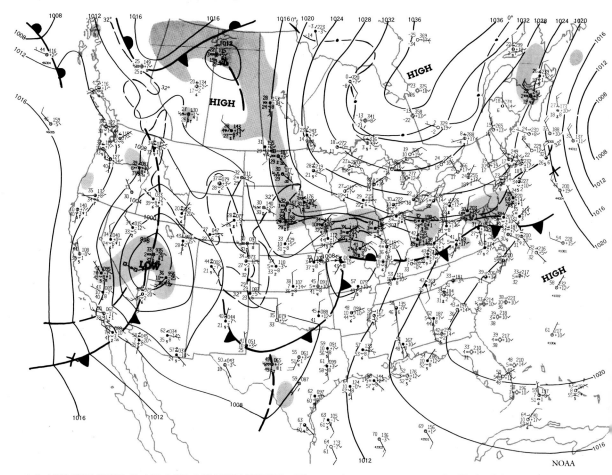

A LATE WINTER DAY ACROSS THE WEST—The weather pattern across the United States on March 15, 1987, is fairly typical for the season, with two low pressure systems dominating the nation's weather. Winds are blowing in a counterclockwise direction around both cyclones. A warm front extends east from a low over Missouri, while a cold front plunges southward behind (west of) the low. The stronger low pressure center over Nevada is trailing a cold front across southern California. The dashed lines extending to the north and southwest of the Nevada low are "troughs"—regions of low pressure and shifting winds that lack strong enough temperature contrasts to qualify as warm or cold fronts. Weak temperature contrasts are common among Great Basin lows and are the hallmark of dissipating storms. This particular storm weakened the next day, but later regenerated into a Great Plains cyclone over the Texas Panhandle. Light snow (denoted by two stars) is falling at Ely, Nevada, just north of the low, while southeast of the low, light rain (two dots) is falling on Cedar City, Utah.

STANDARD WEATHER SYMBOLS

NASA

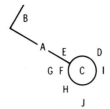

A, B
Direction and Windspeed

Symbol	Speed, knots	Speed, m/sec
◎	Calm	Calm
———	1 - 2	0.5 - 1
⌐	3 - 7	1.5 - 3.6
⌐	8 - 12	4.1 - 6.2
⌐	13 - 17	6.7 - 8.7
⌐	18 - 22	9.3 - 11.3
⌐	23 - 27	11.8 - 13.9
⌐	28 - 32	14.4 - 16.5
⌐	33 - 37	17 - 19
⌐	38 - 42	19.5 - 21.6
⌐	43 - 47	22.1 - 24.2
⌐	48 - 52	24.7 - 26.7
⌐	53 - 57	27.3 - 29.3
⌐	58 - 62	29.8 - 31.9
⌐	63 - 67	32.4 - 34.5
⌐	68 - 72	35 - 37
⌐	73 - 77	37.6 - 39.6
⌐	103 - 107	53 - 55

⟶ (wind direction)

A   Direction from which wind is blowing (see symbols at left)
B   Windspeed (see symbols at left)
C   Extent of cloud cover (see symbols below)
D   Barometric pressure reduced to sea level
E   Air temperature at time of reporting
F   Weather condition at time of reporting (see symbols below)
G   Visibility
H   Dewpoint temperature
I   Pressure change during the 3 hr period preceding observation
J   Height of base of lowest cloud

Missing or unavailable data are indicated by "M" in the proper location.
(Note: Only those codes which appear on maps in this report are listed).

C
Cloud Cover

Symbol	Percent covered
○	Clear
◐	Up to 10
◔	20 to 30
◑	40
◑	50
◕	60
◕	70 to 80
◖	90 or overcast with openings
●	Completely overcast
⊗	Sky obscured

F
Present Weather Conditions

Symbol	Explanation
⌇	Visibility reduced by smoke
∞	Haze
،	Intermittent drizzle (not freezing), slight
••	Continuous rain (not freezing), slight
••	Continuous rain (not freezing), moderate
*	Intermittent snow, slight
▽	Slight rain showers
* ▽	Slight snow showers

around the world is 1,013 millibars, or 29.92 inches. A typical high pressure may be 1,030 millibars (30.42 inches) or more, while an average low has pressures below 1,000 millibars, or 29.53 inches.

Fronts appear on a weather map as heavy lines with little round or pointed bumps. Round bumps mean a warm front, and pointed bumps a cold front. The bumps point in the direction the front is headed. A line with both round and pointed bumps pointing in the same direction is called an "occluded" front. This is where a cold front has caught up to a slower warm front. The passage of an occluded front usually brings rain or snow and a shift of wind direction, but not much of a temperature change. If the line has round bumps on one side and pointed ones on the other, it's a stationary front. Warm or cold fronts may stall and become stationary fronts, while stationary fronts can start moving. Sometimes part of a front is moving as a warm or cold front while another section is stationary. If you follow maps over several days, you'll see fronts form and disintegrate and the temperature contrasts across them strengthen and weaken.

Between all the lines on weather maps, there are little groupings of numbers and symbols. These plots of data from different weather stations are called "station plots." At the center of each station plot is a small circle; the amount of shading in the circle shows the amount of cloudiness. An empty circle means clear skies, while a completely blackened circle means overcast. A line extending from the circle gives the direction the wind is blowing *from*, while the number of branches on the line indicates the wind speed. One branch means 10 knots, or 12 m.p.h., two branches mean 20 knots, and so on. The current temperature appears to the upper left of the circle, and the dew point temperature to the lower left. Symbols between the temperature and dew point mark the current weather. One dot means light rain, two dots mean heavier rain; a six-pointed star means light snow, and so on. Gray shaded areas on the map show where precipitation is falling at the time the map was made. Other numbers and symbols give pressures, cloud types, rainfall amounts, etc. They are all explained in a little pamphlet available from the same folks who publish the weekly booklet of weather maps.

RESOURCES

I hope that by reading this book, you have had the satisfaction of gaining an understanding and appreciation of the special kinds of weather we see in the West. Weather is a fantastically complex subject and, at best, this book has just skimmed the surface. Perhaps it has whetted your appetite for more information. There are many ways to expand your knowledge about weather, ranging from reading books and periodicals to following the daily weather maps and even taking your own weather records. Here are some suggestions for these pursuits.

BOOKS

There are lots of books about the weather, ranging from kids' books and photo albums to textbooks and technical reports. I suggest going to a library and skimming through their selection and finding one that appeals to you. I have a few favorites:

Weather, by Paul Lehr, Will Burnett and Herbert Zim, Golden Press, New York. This pocket-sized volume came out in 1957 and has been updated several times since. It's a compact but comprehensive overview of the weather, easy to read and well-illustrated. I have recommended it to audiences from Cub Scouts to mountaineers.

Weather in the West, by Bette Anderson, American West Publishing Co., Palo Alto, California. Like the volume you have in your hands, this book covers western weather—but in a more descriptive style.

There are a number of good regional weather books written by locals. Here's a partial list; check your library for others.

Montana Weather, edited by Carolyn Cunningham, published by Montana Magazine, Inc., Helena, Montana.

Weather of the Pacific Coast, by Walter Rue, Writing Works, Inc., Mercer Island, Washington. Has a strong emphasis on the Seattle area.

Arizona Climate, by William Sellers and Richard Hill, University of Arizona Press, Tucson, Arizona.

The Weather of the San Francisco Bay Area, by Harold Gilliam, and *Weather of Southern California*, by Harry Bailey, sold by Wind & Weather, Mendocino, California.

Skywatch is the third weather-related book published by Fulcrum, Inc. The previous two are:

The Avalanche Book, by Betsy Armstrong and Knox Williams, and *Acid Rain—Reign of Controversy*, by Archie Kahan.

U.S. GOVERNMENT PUBLICATIONS

Those who relish detailed weather information will find a real gold mine in some of the publications offered by the U. S. Government. Many reasonably priced publications are available from the Superintendent of Documents, U.S. Government Printing Office, Washington, D.C. 20402. Write for their general catalog, a special catalog of weather-related publications, and their monthly listing of new books. They also have walk-in bookstores in Los Angeles, San Francisco, Seattle, Denver, Pueblo (Colorado) and elsewhere in the country.

The National Climate Center, Federal Building, Asheville, North Carolina 28801 publishes tons of climate data for all sorts of locations. Write for their list.

Storm Data describes hundreds of storms, from hurricanes to dust devils, that strike the United States each month. There are maps, photos and statistics—a real bonanza for storm lovers!

Local Climatological Data (your city)—Monthly summaries of the daily weather for dozens of western cities (mostly cities with airports). All the numbers you ever need to know.

Climatological Data (your state)—Monthly summaries of weather at dozens of cities, towns and even remote locations in each state. Includes temperature, rainfall, snow and other data.

Climates of the States (your state)—Published for each state of the Union, these booklets give a narrative description of the climate, several pages of tabulated statistics about temperature, wind, snow and the like for selected cities, and maps of average temperatures and precipitation. Very informative.

Weekly Weather and Crop Bulletin gives a weekly report on weather conditions state-by-state and around the world, with emphasis on effects on agriculture.

PERIODICALS

Weatherwise, published six times a year by Heldref Publications, 4000 Albemarle Street, N.W., Washington, D.C. 20016. For 40 years this has been the only magazine in America devoted solely to weather. Articles cover weather research, history and recent weather events.

Science News, by Science Service, 231 West Center Street, Marion, Ohio 43305. As the name implies, this weekly magazine reports on the latest discoveries in all the sciences, including meteorology.

American Weather Observer is a monthly tabloid that features weather reports from amateur weather watchers around the country. You can even publish your own weather data. Write to *American Weather Observer*, 401 Whitney Boulevard, Belvedere, Illinois 61008.

SATELLITE PHOTOS

Many of you may be impressed by the fine detail of the satellite images of Hurricane Norman in the Hurricanes chapter and of the cloud eddy in the Whirlwinds chapter. These remarkable images were taken by the Air Force's Defense Meteorological Satellite Program (DMSP) satellite. Many more DMSP images are available to the public. Write to the National Snow and Ice Data Center (NSIDC), Campus Box 449, University of Colorado, Boulder, Colorado, 80309.

CLOUD PHOTOGRAPHS

For cloud lovers, the *International Cloud Atlas, Volume II* is a must. Published in 1987 by the World Meteorological Organization, this lavish volume contains over 200 photographs of clouds and other weather phenomena and is *the* international standard for identifying cloud types. Four of the photographs in *Skywatch* are in this atlas. It is available from the American Meteorological Society, 45 Beacon Street, Boston, Massachusetts 02108.

DAILY WEATHER INFORMATION

The ever-changing weather is best appreciated by following it on a daily basis. You don't have to take your own records; there are plenty of data available in the media and elsewhere. Many newspapers carry weather maps and satellite photos, and most have daily high and low temperatures. Most newspaper maps aren't very detailed, and some are just about worthless.

If you want a good weather map, the National Weather Service publishes a weekly booklet with daily maps and upper-air charts. A year's subscription is $60. Write to the Climate Analysis Center, Room 808, World Weather Building, Washington, D.C. 20233.

Television weather broadcasts vary in quality; if the weatherman knows something about the subject, you may learn something by watching. Most show satellite photos of the nation's weather, and some put together time-lapse satellite photographs of moving storms and clouds. These space views of the weather give a perspective on the workings of weather that was unavailable 20 years ago. Watch them for a while and you'll get a real feel for how storms grow, move and die.

The National Weather Service broadcasts its own round-the-clock weather information on a special radio frequency. Special radios are available from Radio Shack and elsewhere for $10 to $60. The regular updates of information right from the horse's mouth are especially interesting when a storm is in progress. The Weather Service also has its own television show. Called "AM Weather," this 15-minute report airs weekday mornings on most public television stations.

WEATHER EQUIPMENT

Some department and hardware stores carry a fair selection of thermometers, rain gauges and other gizmos. If you'd like to see some of the more esoteric equipment, try the catalogs available from:

Weatherwise Books and Instruments, Main Street, New London, New Hampshire 03527;

Science Associates, 31 Airpark Road, Box 230, Princeton, New Jersey 08542;

Edmund Scientific, 101 E. Gloucester Pike, Barrington, New Jersey 08007;

Wind & Weather, Post Office Box 2320, Mendocino, California 95460.

ASK THE EXPERTS

Sometimes you can get literature about weather from the various organizations that deal with the subject. Among these are your local weather service and the agriculture or meteorology departments at colleges and universities. The National Center for Atmospheric Research, Post Office Box 3000, Boulder, Colorado 80303, is in the forefront of weather research, and their Information Office has some interesting public relations blurbs.

Many places have local chapters of the American Meteorological Society. Check for locations with the American Meteorological Society, 45 Beacon Street, Boston, Massachusetts 02108. This is a good way to meet some meteorologists and enjoy interesting presentations.

A FINAL NOTE

Now that you've read several books about weather and subscribe to all the magazines and climate reports and pore over the daily weather map as you listen to Weather Radio and can't walk by your barometer without tapping it, you are a true weather nut. But don't forget to look out your window once in a while. That's where the weather is, and it's putting on a show just for you.

NOTES

[1] Studies of Halley's Comet have revealed that comets are composed largely of water ice, leading some researchers to speculate that the atmosphere may have received its water from space, rather than from within, in the form of falling comets. For the time being, I am sticking with the traditional volcanic theory. However, an interesting point arises: the science of meteorology is named after a Greek word, *meteoros*, meaning "high in the air." Ever since, meteorologists have had to explain that they do not study what we now call meteors (from the same Greek word), those burning bits of space dust often referred to as "shooting stars." Astronomers have long known that this space dust is the remnants of disintegrating comets. So if water vapor, that all-important ingredient of what meteorologists do study, actually does have a common origin with meteors, then the distinction between the terms loses some of its significance.

[2] Allen, Joseph P. 1984. *Entering Space.* Stewart, Tabori and Chang, New York.

[3] Blodget, Lorin. 1857. *Climatology of the United States.* J.B. Lippincott & Company, Philadelphia.

[4] The 134-degree reading has been officially accepted by the National Weather Service as the nation's highest temperature reading. However, there are some doubts about the accuracy of that temperature—the thermometer at Death Valley was having problems at the time. If we choose to discard that reading, and we'll never really know if we should, the next hottest record is 129 degrees, also in Death Valley, on July 18, 1960.

APPENDIX 1. CLIMATIC DATA

One way to appreciate the fantastic variety of climates that are found in the West is to scan this table of climate data for some representative locations. In this listing you will find places 20 miles apart whose climates are as different as Brazil's is from Britain's. It wasn't easy deciding which weather stations, of the thousands in the West, to include in this table, but the 63 stations listed here should be a representative cross section of the West's varied climates. Some of the locations are major urban centers, in or near which many readers might live. Others are popular vacation destinations, ranging from national parks to ski areas. A few are simply places with extreme climates—they're exceptionally hot or cold, snowy or rainy. And one very special place is my backyard weather station in Coal Creek, Colorado. I'll admit to showing off a bit of pride about the climate I live in, but there is another reason for including the Coal Creek station in this table. That is to show that with your own weather station, you can come up with your own climate data. After just a few years, you'll be able to compare your home's climate with that of Death Valley, Pikes Peak or Tombstone, Arizona.

Modern weather stations can produce a bewildering variety of data. The few statistics presented here should give you a fairly complete picture of the local climate. The data included in this listing are:

Elevation—Along with latitude and distance from the ocean, elevation is one of the most important factors in determining a place's climate.

Year Records Begin—The longer weather records have been taken, the more reliable the averages will be. It takes about ten or twenty years to come up with a truly representative average.

Average Temperature—I've found annual average temperatures to be fairly worthless at describing a place's climate—San Francisco and Albuquerque have the same annual average. The average daily high and low temperatures for July and January give a much better picture of the daily and seasonal ranges of temperature. July is the warmest month of the year in most western cities, but along the California coast summer peaks in August or September. So August averages for Los Angeles and San Diego and September averages for Eureka and San Francisco are given instead in the July columns.

Temperature Extremes—These are the highest and lowest temperatures since records began. The longer the period of record, the more extreme the extremes are likely to be.

Average Annual Precipitation and Snowfall—Precipitation includes the water that falls as snow, as well as rain.

Wettest Month—This gives an idea of when the rainy (or snowy) season is.

Average Annual Thunderstorms (Tstm)—This is actually the average number of days with thunderstorms. Some days may have two or more thunderstorms, but this doesn't affect the average. Not all weather stations keep thunderstorm records; for those that don't, the numbers (in italics) were estimated from a map produced by Changery for the Nuclear Regulatory Commission.

Sunshine (sun)—The number here is the percentage of possible sunshine—the annual number of hours that the sun shines, given as a percentage of the annual number of hours there would be if the sky were always clear. Seattlites will be pleased to find that the sun shines on their city nearly half of all daylight hours!

SKYWATCH

CLIMATE DATA

station	elevation	year records begin	average temp July max	July min	January max	January min	temp extremes high	temp extremes low	average annual precip	average annual snow	wettest month	average annual Tstm	sun
Arizona													
Flagstaff	7,006	1898	82	50	42	15	97	-30	20.9	97	Aug	71	77
Grand Canyon													
N. Rim	8,400	1931	77	46	38	20	91	-25	22.8	129	Jan	*60*	
S. Rim	6,950	1916	85	54	41	20	98	-22	14.5	65	Aug	*60*	
Phantom Ranch	2,570	1935	106	78	56	36	120	-9	8.4	0	Aug	*50*	
Phoenix	1,110	1895	105	80	65	39	122	16	7.1	0	Aug	27	85
Tombstone	4,610	1894	94	65	61	34	110	6	12.8	2	July	*80*	
Tuscon	2,584	1900	99	74	64	38	117	6	11.1	1	July	54	86
Yuma	194	1879	107	80	69	43	123	22	2.7	0	Aug	9	90
California													
Death Valley	-178	1912	116	88	65	40	134	15	1.7	0	Feb	*5*	
Eureka	43	1887	62	52	53	41	86	20	38.5	0	Jan	5	50
Los Angeles	270	1878	84	65	67	48	112	28	14.9	0	Jan	6	73
Red Bluff	342	1878	98	66	54	37	121	17	21.5	2	Jan	10	77
San Diego	13	1875	78	67	65	48	111	25	9.3	0	Jan	3	68
San Francisco	52	1875	69	56	56	46	103	27	19.3	0	Jan	2	66
Mt. Shasta	3,535	1914	85	51	42	25	103	-8	37.1	105	Jan	13	
Yosemite	3,970	1905	90	53	47	26	110	-6	36.1	73	Dec	*20*	
Colorado													
Aspen	7,928	1925	80	45	34	8	94	-33	19.4	138	Aug	*60*	
Coal Creek	8,950	1982	72	46	33	14	85	-36	26.1	212	May	82	
Denver	5,283	1871	88	59	43	16	105	-30	15.3	60	May	41	70
Estes Park	7,497	1916	79	46	38	17	98	-39	13.8	37	July	*70*	

station	elevation	year records begin	average temp July max	July min	January max	January min	temp extremes high	extremes low	average annual precip	annual snow	wettest month	average annual Tstm	sun
Grand Jct.	4,855	1892	94	64	36	15	105	-23	8.0	25	Aug	36	70
Lamar	3,617	1903	93	63	45	14	111	-30	14.4	26	July	*48*	
Mesa Verde	7,070	1922	88	58	41	19	100	-20	17.6	80	Aug	*45*	
Pikes Peak	14,134	1874	48	34	9	-4	64	-39	29.7	553	July	*85*	
Idaho Boise	2,838	1900	91	59	37	23	111	-23	11.7	21	Jan	15	64
Pocatello	4,454	1899	89	54	32	15	105	-31	10.9	43	May	23	63
Sandpoint	2,126	1918	83	48	32	19	104	-35	30.6	75	Dec	10	
Sun Valley	5,980	1937	83	38	31	1	96	-46	17.4	122	Dec	*50*	
Montana Billings	3,567	1896	87	58	30	12	112	-49	15.1	57	June	30	60
Glasgow	2,284	1893	84	57	18	-1	113	-59	11.6	28	June	28	
Helena	3,828	1881	84	52	28	8	105	-42	11.4	47	June	35	59
Kalispell	2,965	1897	82	48	27	11	105	-38	15.9	65	June	24	
Nevada Elko	5,050	1910	90	50	37	13	107	-43	9.3	38	Jan	20	
Las Vegas	2,162	1937	105	76	56	33	117	8	4.2	1	Aug	15	85
Reno	4,404	1888	91	48	45	20	106	-19	7.5	25	Jan	13	79
Winnemucca	4,301	1878	93	51	42	17	108	-36	7.8	24	July	14	67
New Mexico Albuquerque	5,311	1893	93	65	47	22	105	-17	8.1	11	Aug	42	76
Carlsbad	3,120	1895	96	68	59	31	112	-7	12.4	5	Oct	39	74
Chaco Canyon	6,175	1951	91	55	43	12	102	-38	8.5	19	Aug	48	
Cimarron	6,540	1903	83	55	47	19	99	-35	15.3	33	July	*110*	

SKYWATCH

station	elevation	year records begin	average temp July max	July min	January max	January min	temp extremes high	temp extremes low	average annual precip	average annual snow	wettest month	average annual Tstm	sun
Sacramento Peak	9,240	1954	72	50	38	21	93	-23	23.5	72	July	62	
Santa Fe	7,200	1876	85	56	42	18	98	-18	13.8	29	Aug	54	74
Oregon													
Astoria	8	1884	69	53	47	35	101	6	69.6	5	Dec	8	
Burns	4,151	1939	84	54	37	18	103	-28	10.1	46	Dec	14	
Crater Lake	6,475	1919	69	41	33	17	100	-21	66.9	541	Dec	12	
Pendleton	1,482	1898	88	59	39	26	119	-28	12.2	18	Jan	10	
Portland	21	1874	80	56	44	34	107	-3	37.4	7	Dec	7	48
South Dakota													
Rapid City	3,162	1900	87	59	32	9	110	-33	16.3	39	June	42	62
Texas													
El Paso			95	70	58	30	112	-8	7.8	6	July	83	
Utah													
Brighton	8,740	1951	73	44	31	8	85	-34	43.1	411	Jan	60	
Bryce Canyon	7,595	1951	80	44	36	5	92	-29	12.1	61	Aug	36	
Canyonlands	4,573	1903	97	62	42	17	113	-24	9.0	12	Oct	35	
Salt Lake City	4,221	1874	93	62	37	20	107	-30	15.3	58	Apr	35	66
Washington													
Quillayute	179	1942	69	50	45	33	99	5	104.5	15	Dec	7	32
Mt. Rainier	5,427	1917	62	43	31	19	92	-20	114.0	578	Dec	15	
Seattle	19	1892	75	56	45	36	100	3	38.8	7	Dec	6	43
Spokane	2,356	1882	84	54	31	20	108	-30	16.7	51	Dec	10	53
Stampede Pass	3,958	1944	66	47	27	19	90	-21	91.1	442	Dec	7	
Wyoming													
Cheyenne	6,126	1871	83	55	37	15	100	-38	13.3	54	May	50	65

station	elevation	year records begin	average temp July max	July min	January max	January min	temp extremes high	temp extremes low	average annual precip	average annual snow	wettest month	average annual Tstm	sun
Jackson	6,230	1917	81	40	26	6	101	-52	15.3	91	May	*50*	
Laramie	7,266	1891	80	48	33	9	94	-50	10.4	48	July	47	
Sundance	4,750	1908	83	55	31	9	102	-39	17.0	69	June	*54*	
Yellowstone (Mammoth Springs)	6,241	1887	78	46	27	9	96	-41	16.5	89	June	*50*	56
CANADA													
Alberta Banff	4,582	1888	72	45	20	2	94	-60	20.7	99	June	6	40
Calgary	3,539	1881	74	49	22	2	97	-49	17.2	61	June	22	50
Edmonton	2,219	1880	74	53	14	-3	99	-57	17.6	52	July	21	51
British Columbia Penticton	1,119	1907	83	53	31	22	105	-17	11.7	27	June	15	45
Vancouver	16	1899	72	55	41	31	92	0	42.1	21	Dec	4	44
Victoria	230	1898	68	52	43	35	95	4	25.9	13	Dec	4	50
Saskatchewan Saskatoon	1,643	1892	79	53	8	-12	104	-55	13.9	44	June	18	55

APPENDIX 2. BUILDING YOUR THERMOMETER SHELTER

When we think of weather, the first thing we think of measuring is temperature. For temperature readings to be of real use, they must be read from a thermometer that is properly protected from sunlight, rain and snow. There are a variety of ways to provide this protection, the most common being to place the thermometer in a louvered wooden box. The following design has served me well for many years.

PARTS NEEDED

The following items are sold in most hardware or lumber stores:

1 - 2-foot x 4-foot x 3/4-inch outdoor-grade plywood

2 - 36-inch-high x 15-inch-wide louvered pine shutters

1 - 3-foot pine 2x4

1 - 7-foot cedar or redwood 4x4

2 - 2-1/2-inch hinges with screws

1 - latch with screws

2 - 5-inch x 1/4-inch lag bolts with washers

50 - 2-1/2-inch nails

1 - pint can exterior-grade white paint

The louvered shutters should have a solid wood crosspiece in the middle of the louvered area. The metal hinges, latch and nails should be zinc-plated to resist rust. The paint pigment should be titanium dioxide, which is very effective in reflecting sunlight. Check the label on the side of the paint can for the listing of ingredients.

TOOLS NEEDED

A hammer, saw, drill, screwdriver, adjustable wrench, paintbrush and shovel.

HOW TO PUT IT TOGETHER

Cut the louvered shutters in half along the solid wood crosspiece. This results in four 15-inch x 18-inch louvered panels, each with solid cross-pieces at the top and bottom. Cut the plywood into one 18-inch x 21-inch and two 15-inch x 18-inch pieces. Cut the 2x4 into one 18-inch length, two 3-inch lengths and two 2-inch lengths.

Now, center one of the 15x18 plywood pieces on the end of the 4x4 beam. Drill two holes through the plywood and into the end of the beam for attaching the two together with the lag bolts (don't attach them yet).

You now make a box, with the drilled 15x18 plywood piece forming the bottom, the other 15x18 plywood forming the top, and the louvered panels forming the sides. Nail the back and side louvered panels to each other and to the bottom pieces, with the back panel fitting "inside" the side panels. The side panels should be about 3/8 of an inch farther apart at the front than at the back. This will allow room for the hinged front panel.

Next, attach the front louvered panel using the hinges and screws. Put the latch on the door. Nail down the top 15x18 plywood piece. Place the 18-inch 2x4 vertically inside the box. Locate it near the center of the box, just behind the two drilled holes in the bottom piece, and nail each end with two nails. This vertical 2x4 will soon be holding your thermometer, so make sure it's firmly in place.

The 18x21 plywood piece forms the roof and shades the box from direct sunlight. Nail the two 3-inch 2x4s on top of the box, near the front, and the two 2-inch 2x4s near the back. Nail the roof to the 2x4 pieces; the roof should slope down toward the back. If you live in a windy location, you may want to beef up the construction with steel angle brackets and screws.

Paint your box, inside and out, with two coats. Dig a two-foot-deep post hole and set the 4x4 beam into it (make sure the end with the drilled holes is up!). Bolt the box to the post, place the thermometer on the 2x4 mounting board, and start reading the temperature.

APPENDIX 3. WIND SPEED SCALES

Wind scale	Wind Speed (mph)	Wind type	Descriptive effects
BEAUFORT			
0	0–1	calm	smoke rises vertically
1	1–3	light wind	smoke drifts slowly
2	4–7	slight breeze	leaves rustle; wind vanes move
3	8–12	gentle breeze	leaves and twigs in motion
4	13–18	moderate breeze	small branches move; raises dust and loose paper
5	19–24	fresh breeze	small trees sway
6	25–31	strong breeze	large branches sway; telephone wires whistle
7	32–38	moderate gale	whole trees in motion; wind affects walking
8	39–46	fresh gale	twigs break off trees
9	47–54	strong gale	branches break; shingles blow from roofs
10	55–63	whole gale	trees snap and uproot; some damage to buildings
11	64–72	storm	some damage to chimneys and television antennas
12	73-82	hurricane	peels shingles from roofs; windows break
FUJITA			
F1	73–112	weak tornado	cars pushed off road; light trailers pushed or overturned
F2	113–157	strong tornado	roofs torn from frame houses; trailer homes destroyed; cars blown from highways
F3	158–206	severe tornado	walls torn from frame houses; cars lifted off ground; trains derailed
F4	207–260	devastating tornado	frame houses reduced to rubble; bark removed from trees; cars and trains thrown or rolled considerable distances
F5	261–318	incredible tornado	Whole frame houses tossed from foundations; cars fly through air; asphalt torn from roads

F1

F2

F3

F4

F5

FUJITA SCALE FOR DAMAGING WINDS

F 1   *(73-112 m.p.h.)*   *moderate damage*

F 2   *(113-157 m.p.h.)*   *considerable damage*

F 3   *(158-206 m.p.h.)*   *severe damage*

F 4   *(207-260 m.p.h.)*   *devastating damage*

F 5   *(261-318 m.p.h.)*   *incredible damage*

Storm Data

INDEX

155

SKYWATCH

upwelling 101, 109

Venus 2, 6, 123, 124
virga 65
volcanoes 2, 5, 121-124

Wasatch Mountains 60, 86, 97, 108, 127
Washoe Zephyr 84, 97
water 4, 5, 10, 24, 26, 31, 73, 76
weather maps 137-140, 142
weather station 99, 129
West Coast 16, 44, 100, 102, 104, 109-114, 127
wind 9, 12-15, 20, 40, 53, 55, 63, 64, 73, 84-97,
        111, 115, 116, 133-135, 137
wind shear 64
wind speed 9, 22, 64, 70, 73, 75, 76, 79-82, 92, 97,
        140, 153, 154

Yellowstone National Park 34, 101, 107
Yuma, Arizona 39, 41, 79, 81, 82, 104, 105, 107